Enhancing Humanity:
An Introductory Look at Issues Surrounding Human Enhancement and Bio-Engineering

Enhancing Humanity: An Introductory Look at Issues Surrounding Human Enhancement and Bio-Engineering

Authors

Austin Mardon
Destiny Lenhardt
Hala Madhi
Syed Rizvi
Hannah Kuipers
Mya George
Paige Breedon
David Henneberg
Iffah Shaikh
Aidan Lang

Editor
Catherine Mardon

GM★ PRESS

First Printing: 2023

Typeset and Cover Design by Monette Rockliff
ISBN: 978-1-77889-037-6
eBook ISBN: 978-1-77889-038-3

Golden Meteorite Press
103 11919 82 St NW
Edmonton, AB T5B 2W3
www.goldenmeteoritepress.com

Table of Contents

Chapter 1: Introduction to Human Enhancement

By Destiny Lenhardt

Looking back one hundred years ago or even a decade ago, we see the strides that humans are making in advancing science and technology. Science is building a foundation for which innovation has a platform to explode in terms of pushing the boundaries of technological capability. This knowledge has incredible implications for enhancing humans' abilities beyond the normal scope. It's no surprise that people have been working continuously and found incredible success in this field. From here, some might think that the only direction is forward. However, these advancements progress non linearly as society navigates the practical and ethical implications on different schools of thought. This book explores this topic and sets out to lay the foundation for understanding what human enhancement is, and what that means in a larger context.

Human enhancement refers to the use of technology, medicine, and other means to improve humans' abilities beyond that which they were born with (Sienna). The term human enhancement does not discriminate on the aspect of the human which is being enhanced. Human enhancements can include many avenues such as surgery or pharmaceuticals in the field of medicine, prosthetics or devices to enhance physical abilities, or the use of brain-computer interfaces or other technologies to enhance cognitive abilities. It can also encompass "low tech" forms such as exercise, diet, education, and other lifestyle choices that are designed to improve human capabilities. As such, human enhancement is a blanket term that envelops most anything relating to positively augmenting one's abilities, even temporarily. Human enhancement has been discussed in various fields. As human enhancement gains attention, it has become the forefront of many conversations in fields including philosophy, psychology, medicine and technology.

The aim of human enhancement is to increase one's abilities in strength, intelligence, memory, or life span by overcoming their physical and (or) mental limitations (Savulescu and Bostrom). Successful human enhancement has improved individual quality of life and humanity as a whole (Masci). The benefits are evident when you see the rising age of life expectancy, the emerging uses of brain-machine interfaces, or even a beautiful sculpture that took years of artistic practice to culminate.

The concept of human enhancement has been present in various forms throughout history. In fact, Masci would argue that the practice is as old as human civilization itself. From the earliest records, human civilizations have been seeking to increase their abilities and out perform in every way. Kiger outlines an early record of human enhancement with the Spartans. In the 6th to the 2nd century BCE, the Spartans worked tirelessly on self improvement through exercise, diet and mental preparation. As the young Spartans grew in age and size, the exercise routine grew as well to ensure continuous gains. If the boys did not meet the requirements, then they received punishment. Difficult as it was, this improved individual strength, fostered community, provided a competitive edge in war, and it influenced the great socio-political stability. For the Spartans, the humans' incredible potential and ability to improve was the driving force in the pursuit of self-improvement that shaped the entire culture of the nation.

While the concept of human enhancement has been around as long as humans themselves, the term "human enhancement" is widely credited to philosopher and bioethicist Julian Savulescu, who first used the term in the early 2000s. Savulescu and Bostrom are prominent voices and writers in this emerging field. It was not until the 20th century that human enhancement became a more widely discussed and researched topic, with the advent of new technologies and scientific breakthroughs in fields such as biotechnology, neurotechnology, and genetics. This has led to the recent developments of a wide range of human enhancement technologies, from drugs and supplements designed to improve cognitive function and physical performance, to bionic limbs and other prosthetics that can replace lost or damaged body parts with functional equivalents. Since the term first started circulating, the topic has been questioned in many fields. Now, human enhancement has been gaining traction exponentially to the point where humanity may be on the cusp of an enhancement revolution (Masci). Today, one of the incredible leaps seen in human enhancement includes a successful implant of an artificial retina into the eye of a blind patient to return partial sight. For these patients, the chance to see loved ones, to gain individual autonomy and feel a new level of independence would likely make it all worthwhile. Human enhancement has been growing for a good reason with all the incredible opportunities it has presented for individuals and humanity itself.

While human enhancement has made an incredible positive impact, there has also been a great number of failures. For example, Day outlines the brutality of a formerly common surgery for improving mental illness, the lobotomy. This surgery had upwards of 15% fatality rate while leaving many others crippled or vegetative for the rest of their life. Fortunately this procedure is not what it used to be. However, today there are still efforts towards human advancements with negative effects. This is often seen for temporary methods of improvement that have long term negative side effects. For example, testosterone is a hormone which can be abused when used as a drug to improve physical appearance and performance (Durbin). It may have short term effects that appear to improve human abilities, however, in the long term, testosterone can have detrimental side effects. Some of the possible negative side effects of misuse include infertility, heart attack, liver disease and addiction. The faults of these technologies bring a reluctance for some in regard to human enhancement. New technologies are emerging quickly, however, with each new advancement comes a list of unknown long term side effects.

Over time, these innovations are optimized and safety is applied at the forefront in an effort to mitigate risk and improve enhancement. What makes human enhancement what it is today is the continual advancement of technology and science, which constantly pushes the boundaries of what is possible. Whether through the development of new drugs and supplements, the growth of the wearable technology market, or the increasing sophistication of prosthetics and bionics, human enhancement remains a dynamic and rapidly evolving field that continues to challenge traditional notions of what it means to be human.

Because human enhancement is so encompassing, it has been categorized into three primary forms, reproductive, physical and mental. As the categories suggest, reproductive enhancements are interventions to support infertility while physical enhancement supports physical attributes and mental enhancement supports cognitive function. The different categories of human enhancement are formed by the system for which they support, not the method that the support is administered. Therefore, a technology that enhances physical and mental abilities can be categorized as both physical and mental enhancement. This also means that the context of the focus changes the type of enhancement. For example, working out is a physical agent that improves physical abilities. However, many people use working out as a treatment for mood disorders or for improving cognitive abilities. In this context, exercise is a form of mental cognitive enhancement. As new technologies emerge, a natural question is, how can this be applied to individuals? It is in that application that attention is given to the relief rather than to the nature of the technology.

Reproductive enhancement refers to the use of assisted reproductive technologies (ARTs) to improve the chances of conceiving a healthy child or to create a child with specific desired traits (Jain and Singh). In the field, subfertility is a term used to describe any form of reduced fertility and prolonged duration prior to conception. Jain and Singh outline many techniques in which obstacles are now overcome for many cases of infertility in both males and females. From an evolutionary standpoint, one's fitness is a representation of their reproductive success and therefore, assisted reproductive technologies are crucial for humanity's wellbeing. From a humanistic standpoint, ARTs are creating families for couples that would have never had the chance to bring a child into the world.

Some examples of ARTS include controlled ovarian stimulation or preimplantation genetic Testing (PGT; Jain and Signh). Instead of utilizing the natural menstrual cycle, controlled ovarian stimulation is now available to maximize the number of oocytes gained and decrease the period prior to conception. PGT can be used to select embryos that are free of genetic disorders or even to choose the sex of the embryo. Through combining these techniques, in vitro fertilization can be used to transfer a viable embryo into the uterus. In vitro fertilization with gamete donation, can also increase the health of a fetus or even impregnate women outside of heterosexual relations.

In vitro fertilization was developed in the 1970s and the first successful birth was in the United Kingdom on July 25, 1978 (Manganaro; Jain and Singh). The procedure was developed by Dr. Robert Edwards and Dr. Patrick Steptoe, who had been working on the technique for over a decade. From this work, Edwards later received a share of the 2010 Nobel Prize for Physiology or Medicine. After timing the procedure with natural ovulation, Edwards and Steptoe collected the eggs from the ovaries using a needle, and fertilized them with sperm in a laboratory dish. The fertilized eggs were then incubated for a few days until they developed into embryos, which were implanted into the uterus of Lesley Brown. Lesley's duration to conception was over nine years due to her blocked fallopian tubes. Louise Brown was the first 'test-tube baby' born, however, currently there are roughly five million babies globally conceived from ARTs. This is a clear example of the development and application of reproductive human enhancement.

The category of physical human enhancement refers to the use of technology or medicine to improve physical performance and (or) appearance. The previous example from Durbin on physical enhancement via the use of anabolic steroids by athletes would be categorized as physical enhancement. This is because anabolic steroids are synthetic hormones that mimic the effects of testosterone to promote physical enhancement such as improving muscle growth, strength, and endurance. The use of steroids by athletes became widespread in the 1950s and 1960s, and their use continues today. Another example, Solomon et al. outlines a form of physical enhancement called 'gene-doping'. This is the non-therapeutic use of genes as a performance-enhancing drug (PEDs). With the pressure of competition in athletes and even beauty related social norms, the use of these genetic elements have become more common. In this regard, other examples of physical enhancement include cosmetic surgery, such as breast augmentation or liposuction. Some of these invasive plastic surgeries are very costly and come with side effects. However, some are medically based, less severe and have benefits including decreased pain and increased quality of life (Furnham and Levitas). Further examples of physical enhancement include prosthetics, such as bionic limbs, hearing aids and the former case from Masci of successfully implanting artificial retinas into the eyes of blind patients.

Medical and technical advancements have changed the scope of physical enhancement, however, it is not a new concept. The desire to enhance physical attributes has been present in humans throughout history. Grivetti elaborates on one of the earliest examples of physical enhancement, using ergogenic aids. Ergogenic aids were used by ancient Greek athletes to improve their appearance and performance in training. Ergogenic aids include many specific substances or techniques that act as performance enhancers. The ancient Greeks used a variety of these aids, including herbs, wine, and animal testicles, for their athletic competitions.

Additionally, an early example of physical enhancement was the use of prosthetics in ancient civilizations, such as the iron hand used by the Roman general, Marcus Sergius (Olson and Zuo). These prosthetics allowed individuals with amputations or injuries to regain some physical function and mobility, enhancing their performance of daily activities or even engagement in combat.

Overall, physical enhancement has a long and complex history, with various substances, techniques, and technologies throughout different time periods and cultures. However, throughout this evolution, something has remained constant. People are still motivated to improve their physical appearance and athletic performance. The growth of globalization has increased the reach of national sports, changed beauty standards and normalized the extremes. This increased social pressure perpetuates the need for further enhancement. If the goal of human enhancement is to improve abilities beyond what is normal, then normalizing these extremes will make it a necessity to keep up rather than an advancement to get ahead.

The third category of human enhancement is mental. Mental or cognitive enhancement refers to the use of technology or medicine to improve cognitive function, such as memory, attention, or creativity. Mental enhancement includes a variety of different cognitive enhancement strategies. For example, biochemical agents include the use of nootropics and even the application of ordinary substances such as oxygen, which are also known as "smart drugs". Physical forms of mental enhancement include neurofeedback, which uses EEG technology to measure brain waves and provide real-time feedback to help individuals train their brains to improve specific cognitive abilities. Lastly, there are behavioral methods such as meditation for improving cognitive abilities (Dresler et al.).

Various substances and techniques have been used throughout history to enhance one's cognitive function. Kapalka outlines the cognitive benefits from an early example of mental enhancement. This was the use of herbal remedies in traditional Chinese medicine, which included ginseng and ginkgo biloba. Ginkgo biloba was used to commonly treat memory loss and is still used today for improving memory and even elevating mood.

Mental enhancement has continued to gain attention in the public eye. The 1960s gave rise to nootropic substances like piracetam which was found to improve memory and learning in laboratory animals. Subsequent research showed that it could also enhance cognitive function in humans. The term "nootropic" was coined by Dr. Corneliu Giurgea, a Romanian chemist. The word nootropic is derived from the Greek words noos (mind) and tropein (toward), meaning drugs that act on the mind. Giurgea used this term to categorize piracetam because it did not fit any of the known groups of psychotropic drugs in pharmacology at the time (Poschel).

Today there are various agents of mental enhancement that are used daily for improving cognitive function such as caffeine, Adderall, and Modafinil. These substances can be used as a treatment for disorders like attention deficit hyperactivity disorder (ADHD) but they have also been used by healthy individuals to enhance cognitive performance beyond the normal scope (Dresler et al.). Overall, the history of mental enhancement is complex and multifaceted, with various substances and techniques used to improve cognitive function. Throughout the history of mental enhancement and today, techniques and technologies have been developed to overcome and enhance many aspects of cognition.

Many of the examples listed here outline the vast research on human enhancement which is characterized by a wide range of technologies and approaches. The current position of human enhancement is seen on the upward. This is in part due to the advancement of other fields such as biotechnology, nanotechnology, and artificial intelligence. Scientists are seeking ways to converge these innovations and integrate them for improving human abilities.

Looking to the future, there are a number of potential directions that human enhancement research could take. Currently, there are many large goals for each of the categories of human enhancement that scientists are working towards. For reproductive human enhancement, professionals are researching full genetic editing to remove the unknowns of reproduction and have input on the health, appearance and even gender of babies. In physical human enhancement, innovation is working towards a robotic exoskeleton with built in AI to increase physical abilities and labor work efficiency. For mental human enhancement, people are excited by brain-machine interface and there is motivation to apply it with brain computer chip implants to improve cognitive abilities (Rainie et al.). Advancements that started as an inconceivable thought are now available to the public. In a few years, the landscape of the field will flip as new discoveries are made but the trajectory is likely to continue forward.

Despite this progression, the rate of change for the field of human enhancement is not simply parallel to the rate of growth in research and innovation. As new technologies emerge, they must be tested, marketed and accepted by the users. If human enhancement has the goal of improving human wellbeing, then it is critical to ensure that the integration will meet that. Improvements in one area, with costs in another, may not be worth the benefit. As previously discussed, Durbin gave a concrete example of the negative side effects from excessive use of PEDs. Today there are many bodies regulating the availability of human enhancements options to ensure safety and efficacy. Sporting corporations, governments, Food and Drug Administration (FDA), and World Health Organization (WHO) are among some of the most prominent. While human enhancement has made incredible progress and helped billions of people throughout history, there are legal and ethical implications for people to consider throughout this process. While some people embrace the change wholeheartedly, there are many contexts for which this must be considered. If too much change happens too fast, there could be substantial negative by-products.

Rainie et al., outlines the current perspective on human enhancement which ultimately is met with a lot of hesitation because of the unknown. Common concerns for Americans include decreased autonomy, and increased economic disparity. The research also showed that many people are unfamiliar with the topic and unable to make informed decisions.

There has been an increase in educational resources, however the media is poorly regulated and can be providing some false information. It is not surprising that many people are on the fence about welcoming advancements in human enhancement.

The progression of human enhancement is seemingly inevitable. Having an understanding of the topic makes you a stakeholder, increases your autonomy and positions you at the forefront of innovation. In the following chapters we will look at the current landscape of information and provide a foundation for which people can make clear judgments about human enhancement.

References

Day, Elizabeth. "He Was Bad, So They Put an Ice Pick in His Brain..." The Guardian, Guardian News and Media, 13 Jan. 2008, https://www.theguardian.com/science/2008/jan/13/neuroscience. medicalscience#:~:text=Despite%20a%2014%20per%20cent,in%20 a%20persistent%20vegetative%20state.

Dresler, Martin, et al. "Hacking the Brain: Dimensions of Cognitive Enhancement." ACS Chemical Neuroscience, U.S. National Library of Medicine, 20 Mar. 2019, https://www.ncbi.nlm.nih.gov/pmc/articles/ PMC6429408/.

Durbin, Kaci. "Testosterone Injections: Uses, Side Effects & Warnings." Drugs.com, 28 Apr. 2022, https://www.drugs.com/testosterone.html.

Furnham, Adrian, and James Levitas. "Factors That Motivate People to Undergo Cosmetic Surgery." The Canadian Journal of Plastic Surgery = Journal Canadien De Chirurgie Plastique, U.S. National Library of Medicine, 2012, https://www.ncbi.nlm.nih.gov/pmc/articles/PMC3513261/.

Grivetti, Louis. Diet, Training, and Ergogenic AIDS: Search for the Competitive Edge ... VIIth IOC Olympic World Congress on Sport Sciences Physical, Nutritional, and Psychological Care of the Athlete in the 21st Century, Oct. 2003.

"Human Enhancement." SIENNA, https://www.sienna-project.eu/enhancement/.

Jain, Meaghan, and Manvinder Singh. "Assisted Reproductive Technology (ART) Techniques - Statpearls - NCBI ..." National Library of Medicine, 28 Nov. 2022, https://www.ncbi.nlm.nih.gov/books/NBK576409/.

Kapalka, George M. Nutritional and Herbal Therapies for Children and Adolescents a Handbook for Mental Health Clinicians. Academic Press, 2010.

Kiger, Patrick. " How Ancient Sparta's Harsh Military System Trained Boys Into Fierce Warriors." History, 8 Sept. 2020, https://www.history.com/news/sparta-warriors-training.

Manganaro, Christine. "Louise Brown." Encyclopædia Britannica, 21 July 2022, https://www.britannica.com/biography/Louise-Brown.

Masci, David. "Human Enhancement." Pew Research Center Science & Society, Pew Research Center, 13 Apr. 2022, https://www.pewresearch.org/science/2016/07/26/human-enhancement-the-scientific-and-ethical-dimensions-of-striving-for-perfection/.

Olson, Jaret, and Kevin Zuo. The Evolution of Functional Hand Replacement: From Iron Prostheses to Hand Transplantation. U.S. National Library of Medicine, 2014, https://pubmed.ncbi.nlm.nih.gov/25152647/.

Poschel, B. P. "New Pharmacological Perspectives on Nootropic Drugs." Handbook of Psychopharmacology, 1988, pp. 437–469., https://doi.org/10.1007/978-1-4613-0933-8_11.

Rainie, Lee, et al. "5. What Americans Think about Possibilities Ahead for Human Enhancement." Pew Research Center: Internet, Science & Tech, 17 Mar. 2022, https://www.pewresearch.org/internet/2022/03/17/what-americans-think-about-possibilities-ahead-for-human-enhancement/.

Savulescu, Julian, and Nick Bostrom. Human Enhancement. Oxford University Press, 2013.

Solomon, Louis M., et al. "Physical Enhancement of Human Performance: Is Law Keeping Pace with Science?" Gender Medicine, vol. 6, no. 1, 2009, pp. 249–258., https://doi.org/10.1016/j.genm.2009.04.008.

Chapter 2: Technological Advances in Human Enhancement

By Syed Rizvi

Introduction

Throughout history, technology has been a marker for human progress and advancement. However, in recent decades, technology has taken huge leaps in revolutionizing the way we live our daily lives. The rate of scientific research is growing at an exponential rate, and has opened many different avenues for human enhancement. With the advancement in the field of artificial intelligence, such as the very recent AI called "ChatGPT", the common conception of what machines are capable of doing is changing. Furthermore with the growing research in the field of genetic engineering, gene therapy, and neurotechnologies, the line between man and machine is diminishing. Humans will now be able to enhance their cognitive, sensory, and physical abilities with the help of new technologies, paving the way for a future where the limits of human potential are constantly challenged. Whether it is virtual reality, brain to computer interface chips, gene manipulation in human DNA, or increased cognitive abilities through neurotechnology, current society will be seen as the age of pioneers again. However, there are many concerns that emerge from these new fields of studies. Many scientists are worried about the effects that humanity can have if these technologies are commercialized. For instance, BCIs are an incredible neurotechnology that allows a computer to interfere with brain signals in order to complete a specific task. However, with the great advancements it can bring in human enhancement, BCIs also bring up certain moral and social issues. For instance, there are worries regarding the confidentiality and security of brain data as well as the possibility that BCIs could be used for non-medical objectives like mind control or monitoring. To guarantee that BCIs are used in a way that is safe and ethical, as with any new technology, it is crucial to carefully analyze these challenges and to set suitable standards and laws. As we continue to explore the frontiers of technology in human enhancement, such as genetic engineering, biotechnology, and neurotechnology, one thing is certain: the impact it will have on our lives, our communities, preexisting justice systems, government and our world will be profound and revolutionary.

Genetic Engineering

Genetic engineering and gene therapy are two closely linked technologies that involve the alteration of genetic materials within an organism. The distinction between the two is based upon purpose. While genetic engineering focuses on modifying the genes to enhance the organism's capabilities beyond its normal means, gene therapy aims to correct genetic defects and/or deter genetic diseases. The discovery of the structure of DNA researched by James Watson and Francis Crick at Cambridge University in 1953, catapulted the progress of genetic development at a molecular level (Howles). However, in 1960, the discovery of an enzyme called DNA ligase, a glue that keeps the DNA strand together, and the isolation of the first restriction enzyme using molecular scissors that cut DNA at precise sequences, allowed the resurgence of research on gene manipulation (Howles). Furthermore in the 1972s, the first recombinant DNA molecules were generated at Stanford University. The use of recombinant molecules is vast and developed the field of genetic engineering. Recombinant molecules are used to sequence the passenger DNA in order to work out the order of successive bases. Moreover, the function of genes can be deduced by expressing the DNA sequence of a gene in a cell, whereas then it can be introduced into an organism that is lacking or deficient. Cloned DNA sequence can be used as a diagnostic tool and passenger DNA can be used to synthesize a protein of interest (Howles). The discovery of recombinant molecules introduced with bacterial cells had the ability to replicate; later coined the term - molecular cloning of gene cloning (Howles). The progress in the genetics field is heavily dependent on the "hard" and "soft" technology available during its time; such as laboratory and computing technologies, and enzymes and biological supplies (Nicholl). Around the time that gene cloning was permissible, gene cloning allowed the isolation of specific DNA sequences, whereby the structures of said gene could be studied on a molecular level. Furthermore DNA sequencing with the help of new automations made the unfathomable task possible - determining the sequence of the human genome (Nicholl). In order to truly understand the basis of genetic engineering, one must understand how little genetic information is required in order to enable a cell to carry out a task. To simplify a very complicated procedure, life is essentially directed by four nitrogenous bases: adenine (A), guanine (G), cytosine (C), and thymine (T) (Nicholl). Enzymes catalyze the reactions of metabolism, in order to express genetic information. Thus, proteins are "condensation heteropolymers synthesized from amino acids, of which 20 are used in natural proteins. Thus, the diversity of protein form and functions is used by a system of sets of three nucleotides, or codons. Genetic engineering still requires a variety of additional research before researchers can attempt gene manipulation within the human domain, however, it has been attempted on livestock and plants. Gene manipulation is a success with potato productions, focusing on crop enhancement (Saeed et al.). *Solanum tuberosum L.* (Potato) is one of the most edible crops around the globe, with more than 5000 types of potatoes (Saeed et al.). Genetic engineering improves potato production largely, due to introduction foreign genes into the nucleus or chloroplast of the crop, in order to improve resistance to abiotic and biotic stresses, such as insect pests, cold, drought, salt, and oxidative stresses. Furthermore, the transplastomic strategy offers numerous benefits for enhancing potato production, such as the capability to produce high and long-lasting proteins, containing biological transgenes, avoiding gene silencing and position effects, and enabling the coexpression of multiple genes. These advantages make transplastomic technology a valuable tool in improving potatoes by promoting the production of insecticidal, antibacterial, and antifungal compounds to enhance their quality and yield (Saeed et al.). Genetic engineering of a potato turns it into a very important crop, where it helps human enhancement by addressing the challenges of increasing population and food security. It is an optimistic outlook to wonder about the several applications genetic engineering can have if implemented within the human domain, however, there is not enough studies to calm the crowds of people who are afraid of the use of technology. The fear mostly stems from humans having the power to "tweak" or change the order which nature has already set itself in. One example of using genetic engineering within the human domain is a designer baby. A designer baby is a baby originated from embryos created by in-vitro fertilization (IVF) . As technology progresses, scientists are able to create designer babies by actively removing disease genes from carrier couples from the carrier embryos (Pang and Ho). In the UK, the approval of mitochondrial DNA replacement therapy has introduced many debates in the genetic engineering realm. The method is controversial due to the alteration becoming inheritable (Pang and Ho). Furthermore, many scientists are pessimistic about advancing technologies due to the power given to humans. The technologies start to attempt to remove diseases and can potentially turn into a means of improving aesthetics. A unique version of eugenics can occur if these technologies are not supervised carefully. Regardless, researchers are making significant advances in the field of genetic engineering, however the long term studies on certain procedures, such as mitochondria DNA replacement therapy, are still uncertain.

Gene Therapy

Gene therapy is a form of molecular medicine with revolutionary potential to significantly enhance human health (Verma and Weitzman). Gene therapy is essentially simple: introduce a piece of genetic material into a target cell which will result in the cure or deter of the disease. In order to gene transfer into cells, tissues, and organs, expensive and complex technologies are required. The process of delivering genetic materials is called transduction. Successful transduction requires crossing a variety of obstacles common to vector systems (Verma and Weitzman). Firstly, the vector should be capable of targeting the cell type most relevant to the disease, whether it includes dividing the targeted cell or non dividing. The transduction process must not fail during the intracellular traffic, vector uptake, and gene regulation (Verma and Weitzman). It is desirable to develop stable gene expression which integrates the vector DNA into the hoist DNA (Verma and Weitzman). Although gene therapy is a revolutionary potential treatment for many genetic disorders, cancer is the most common disease gene therapy is being used for (Arabi et al.). CAR T cells are one of the most successful gene therapy applications in immunotherapy. In order to react against tumor associated antigens, T cells are engineered as chimeric receptors to generate CAR T cells (Arabi et al.). The combination of "T cell activation and expansion ability with scFv specificity makes CAR T cells one of the most promising cancer-fighting tools" (Arabi et al.). Several gene therapy products are approved for treatment of cancer, including gendicin, oncorine, rexin-G, imlygic, delytact and eight CAR T cells - "including kymriah, yescarta, tecartus, breyanzi, abecma, ARI-0001, carteyva and carvykti" (Arabi et al.). Furthermore, gene therapy is not only covered in the domain of cancer research but also with different genetic disorders. The National Human Genome Research defines genetic disorders as "a disease caused in whole or in part by a change in the DNA sequence away from the normal sequence". Many of the clinical trials using gene therapy focuses on metabolic disorders, blood coagulation disorders, and eye disorders (Arabi et al.). Genetic disorders are mainly divided into three types - monogenic disorders, chromosomal disorders, and multifactorial disorders. Monogenic disorders are caused by a mutation in one gene with clear inheritance patterns (autosomal dominant/recessive, X/Y linked, and mitochondrial) (Jiang et al.). One of the more common monogenic disorders is hereditary hearing loss, familial hypercholesterolemia, sickle cell anemia, and cystic fibrosis (Jiang et al.). Hereditary hearing loss contains all forms of inheritance patterns and is a great application for gene therapy. Hereditary hearing loss may be treated using gene therapy, according to research. In this method, the cells of the inner ear are given functional copies of the genes that cause hearing loss. It is believed that by reactivating these genes, hearing loss can be avoided or perhaps reversed.Clinical treatment for hereditary hearing loss are cochlear implants or hearing aids (Jiang et al.). The deafness gene - TMCI, is the most popular target for gene therapy. However, despite the treatment, many cell types in mammalian inner ears lack the ability for regeneration and are considered to be terminally differentiated (Jiang et al.). Therefore, in theory, the therapeutic effects of gene therapy in the inner ear should be long-lasting. However, practically, the duration of gene therapy differs greatly. The reducing efficacy of gene therapy is linked with the generation of neutralizing antibodies fighting against the viral vectors and exogenous genes. Thus, despite striving far in gene therapy models, researchers still need a better alternative to have longer and stable therapy treatments for hereditary hearing loss. Even though gene therapy for genetic hearing loss is still in its infancy, preliminary studies have yielded encouraging outcomes. If successful, this strategy could provide hereditary hearing loss sufferers with a novel and potent therapy alternative.

Neurotechnology

Neurotechnology is a rapidly developing field that has the potential to revolutionize the way we interact with our brains and the world around us. This technology has many uses for human enhancement such as human abilities, including memory, attention, learning, and communication. However, there are many other technological, social, ethical, and environmental factors that need to be taken into consideration. One of the great areas of neurotechnology is brain-computer interfaces (BCIs). Recent advances in the field of neuroscience has made the impossible possible by allowing the record of large assemblies of neurons and decoding their activities to extract information (Roelfsema et al.). The extraction of information allows different technologies to interfere and stimulate the brain in a desired manner. Currently, many technologies are restricted to therapeutic context, however, as research continues, this may no longer be the case. As technologies progress, it is becoming more likely to record activities of many nerve cells using invasive and non-invasive methods. Some non-invasive methods involve measuring brain activities using electroencephalography (EEG), functional magnetic resonance imaging (fMRI), and infrared spectroscopy (Roelfsema et al.) For example, patient that were thought to be in a vegetative or minimally conscious state were given instructions, and "asked to imagine playing tennis, they activated their supplementary motor area (SMA), whereas their parahippocampal place area (PPA) was activated when they were asked to imagine walking around in their house (Roelfsema et al.). However, despite being able to predict brain thoughts using these technologies, an obvious limitation is that they have to be within scanners. There are several known EEG signals that are known to control brain-computer interfaces, however, they can be divided into endogenous and exogenous signals (Velasco-Álvarez et al.). "Endogenous signals are based on the self regulation of the EEG in the absence of external stimuli, and the most commonly used are sensorimotor rhythms (SMR) elicited by motor imagery (MI) tasks. Exogenous signals are elicited by an external stimulus; the most frequently used are steady-state evoked potentials (SSEPs) and event-related potentials (ERPs)" (Velasco-Álvarez et al.). However, in recent research in neurotechnology, invasive measures are put with the need of a brain surgery (Roelfsema et al.). For instance, upon the surgery, there is a brain-computer interface that can be of assistance for people with paralysis. By implanting a subdural electrode over the cortex, electrodes can provide low bandwidth of the local EEG, whereas high bandwidths can be obtained via electrodes arrays that are inserted in the cortex. It records neuronal spiking activity and can decode one's intention to move an arm or hand in a particular direction (Roelfsema et al.). There are several potential uses for BCIs, especially in the field of medicine. For those who are paralyzed or have other motor limitations, BCIs can be utilized to help them control prosthetic limbs or other assistive devices. A variety of neurological diseases, including Parkinson's disease and epilepsy, can also be treated with them. BCIs may potentially find utility in the gaming and entertainment industries, where they might be utilized to build immersive experiences that react to the user's feelings and ideas. BCIs can also be utilized to track cognitive performance and provide feedback for training and learning improvements. Furthermore, in 1998, Warwick's chip implants that were inside his hands allowed him to open doors, turn on lights, and control other devices without any physical movement or touching (Teunisse et al.). In the following years, many advances were made and chip implants integrated into the nervous system, starting cyborg communities. Many propose the idea of creating cyborg humans in order to strive for exponentially great enhancement, however many are against the mission, stating that the long term studies of these technologies have not been carefully tested. For example, during a Ted talk, electrodes were applied to two different participants' arms. One of them was able to move the other's arm using the electrical current of the first person's arm, and applying a correlated electrical current to the latter; causing the second participant's arm to move (Teunisse et al.).

Conclusion

Technology has improved quickly in recent years, opening up a variety of new opportunities for improving human performance. Technology is rapidly being used to improve human performance, from prosthetic limbs and brain-computer interfaces to genetic engineering and artificial intelligence. These developments have the potential to greatly enhance our quality of life, but they also bring up significant moral and societal issues that demand attention. The possibility of exacerbated socioeconomic inequality is one of the most important worries about human enhancement technologies. It's feasible that only those who can afford to pay for expensive modifications will have access to them, further separating the wealthy from the rest of society. Concerns exist over the pressure that could be put on people to improve themselves in order to be competitive in the work market or other aspects of life. Based on the degree of one's upgrades, this can lead to the emergence of a new type of social hierarchy. Human enhancement technologies provide a lot of potential advantages despite these worries. For instance, assistive technologies like prosthetic limbs can significantly raise the quality of life for people with disabilities. Genetic engineering could be used to stop genetic diseases and problems from developing before birth, and brain-computer interfaces have the potential to treat a variety of neurological conditions. Human improvements could be utilized to boost cognitive skills like memory and concentration in addition to these medicinal purposes. This might have a big impact on education and training, as well as people looking to develop their own skills in different fields. Controlling and regulating human enhancement technology is one method to allay these worries. This could entail defining standards for the creation and use of these technologies as well as developing systems for tracking their effects on society. It is crucial to include a wide range of stakeholders, including ethicists, decision-makers, and people of impacted communities, in debates about human improvement. The need to make sure that human enhancements are created in a method that is safe, efficient, and available to everyone who needs them is another crucial factor. This can entail funding the advancement of these technologies through research and development as well as assuring their accessibility and affordability. Furthermore, it brings upon dillimas not yet visible due to our current capacity of knowledge and awareness. As technology improves, there will be different advantages and disadvantages that will show up. In conclusion, as one of the classic quotes from the Spiderman series goes, "With great power, comes great responsibility". As a society, we have to govern and practice safe usage of these technologies in order to effectively use it for human enhancement. By working together, we can make sure to address issues to ensure that these amazing technologies are used in a manner that maximizes the benefit while minimizing the risk for humanity.

References

Almeida, Mara, and Rui Diogo. "Human Enhancement." Evolution, Medicine, and Public Health, vol. 2019, no. 1, Jan. 2019, pp. 183–89, https://doi.org/10.1093/emph/eoz026. Accessed 15 Feb. 2023.

Arabi, Fatemeh, et al. "Gene Therapy Clinical Trials, Where Do We Go? An Overview." Biomedicine & Pharmacotherapy, vol. 153, Sept. 2022, p. 113324, https://doi.org/10.1016/j.biopha.2022.113324. Accessed 15 Feb. 2023.

Howles, C. "Genetic Engineering of Human FSH (Gonal-F(R))." Human Reproduction Update, vol. 2, no. 2, Mar. 1996, pp. 172–91, https://doi.org/10.1093/humupd/2.2.172. Accessed 5 Nov. 2019.

Jiang, Luoying, et al. "Advances in Gene Therapy Hold Promise for Treating Hereditary Hearing Loss." Molecular Therapy, Feb. 2023, https://doi.org/10.1016/j.ymthe.2023.02.001. Accessed 25 Feb. 2023.

Nicholl, Desmond S. T. An Introduction to Genetic Engineering. Cambridge University Press, 2003.

Old, R. W., and S. B. Primrose. Principles of Gene Manipulation. Univ of California Press, 1981.

Pang, Ronald T. K., and P. C. Ho. "Designer Babies." Obstetrics, Gynaecology & Reproductive Medicine, vol. 26, no. 2, Feb. 2016, pp. 59–60, https://doi.org/10.1016/j.ogrm.2015.11.011. Accessed 25 Feb. 2023.

Roelfsema, Pieter R., et al. "Mind Reading and Writing: The Future of Neurotechnology." Trends in Cognitive Sciences, vol. 22, no. 7, July 2018, pp. 598–610, https://doi.org/10.1016/j.tics.2018.04.001. Accessed 25 Feb. 2023.

Saeed, Faisal, et al. "Role of Genetic Engineering in Improving Potato Production." Potato Production Worldwide, 2023, pp. 303–15, https://doi.org/10.1016/b978-0-12-822925-5.00006-2. Accessed 25 Feb. 2023.

Teunisse, Wessel, et al. "Human Enhancement through the Lens of Experimental and Speculative Neurotechnologies." Human Behavior and Emerging Technologies, vol. 1, no. 4, Oct. 2019, pp. 361–72, https://doi.org/10.1002/hbe2.179. Accessed 25 Feb. 2023.

Vedder, Anton, and Laura Klaming. "Human Enhancement for the Common Good—Using Neurotechnologies to Improve Eyewitness Memory." AJOB Neuroscience, 2014, www.tandfonline.com/doi/full/10.1080/21507740.2010.483996. Accessed 25 Feb. 2023.

Velasco-Álvarez, Francisco, et al. "Brain-Computer Interface (BCI)-Generated Speech to Control Domotic Devices." Neurocomputing, vol. 509, Oct. 2022, pp. 121–36, https://doi.org/10.1016/j.neucom.2022.08.068. Accessed 25 Feb. 2023.

Verma, Inder M., and Matthew D. Weitzman. "GENE THERAPY: Twenty-First Century Medicine." Annual Review of Biochemistry, vol. 74, no. 1, June 2005, pp. 711–38, https://doi.org/10.1146/annurev.biochem.74.050304.091637.

Chapter 3: The Ethics & Morality of Human Enhancement

By Hala Madhi

Introduction

The ethics and morality of human enhancement has been a widely debated topic for as long as human enhancement itself has been practiced. These implications of human enhancements are incredibly complex and nuanced, and there is certainly no 'one size fits all' answer for whether human enhancement is or is not ethical or moral. In this chapter, the ethics and morality of this subject will be discussed, starting with a differentiation between the two and a discussion on some of the nuances of ethics and morality of human enhancement. Then, an overview of the following ethical considerations relating to this topic will be given: freedom and autonomy, fairness and equity, and human dignity and identity. Finally, the chapter will be completed by presenting a case study of a human enhancement and presenting some of the ethical considerations one must examine when appraising the situation.

The Difference Between Morality and Ethics in the Realm of Human Enhancement

To have a discussion on the ethics and morality of human enhancement, one must first establish a clear framework differentiating the two, as they are often conflated within the subject. One way of thinking about it is that oftentimes ethical issues are discussed in the context of a specific practice/profession, such as ethical code of those in medical professions (Miah). Morality on the other hand can be seen as relating to broader questions or concerns where there might not be 'formal' codes to abide by, such as having a general moral concern about a society with medically enhanced individuals (Miah). It can also be useful to think about it in simpler terms, as in both ethics and morality are concerned with differentiating between "good and bad" or "right and wrong", but ethics comprises the values of a community/group, while morality is more personal (Miah). Therefore, ethics often tends to be more formal, with clearer expectations and delineations, where morality is more normative and can often be difficult to offer an undisputed answer to a question of morality. This is especially helpful to understand when discussing human enhancements, because while certain issues are certainly ethical in nature, such as those concerning physicians who must abide by a very clear-cut ethical code, other concerns are morality-based and more difficult to derive conclusions about. An interesting thought experiment is considering a situation whereby an intervention is declared unethical, and therefore illegal, for physicians to administer, but the actual intervention is not declared illegal (Juengst and Moseley). This implies that while ethically, physicians are not permitted to dispense it, other avenues of procuring these interventions are permitted and therefore may be considered 'moral' by those who choose to pursue them (Juengst and Moseley).

The Nuances of Ethics and Morality in the Context of Human Enhancements

The ethical concerns relating to human enhancements are incredibly varied, which is understandable considering all the possible forms of human enhancement. As discussed in the previous chapter, the field of human enhancement includes both using natural methods (i.e physical exercise), as well as eugenics, so it is imperative to be critical and nuanced when discussing the ethics of the matter. It is also important to consider the stakeholders involved when discussing ethics, as they can vary from very specific individuals or groups to very broad and all-encompassing (Miah). An example would be how one physician's ethical dilemma regarding a specific enhancement intervention (i.e administering growth hormone to a below-average height young boy) would be a very specific ethical decision to be made following the physician's professional code of ethics. However, the issue of genetic modifications is one that is much more broad with many stakeholders involved, in fact one could even argue that the entire world's population are stakeholders considering the far-reaching implications of the concern (Miah). It is also interesting to consider future generations as potential stakeholders whose interests must be considered when making ethical and moral decisions regarding this field. Oftentimes with massive technological and medical advancements such as those in the field of human enhancements, the impact goes further than those directly impacted by the first administrations of the enhancements. Setting precedent for these interventions greatly impacts future generations, and it can be a question of autonomy when it comes to whether it is ethical or moral to go through with these advancements without the awareness and consent of those incoming generations who will inevitably be impacted by them (Miah).

Morality and ethics are also often context specific and change over time. An egalitarian approach to the ethics of human enhancements as a result can be quite complicated and not always morally 'correct' (Allhof and Lin). For example, in technical terms, vaccinations may be considered human enhancement, as they are adding upon the 'natural' state of a human. With an egalitarian approach, it would not be ethical to prohibit one individual or a group of individuals from being vaccinated when it is available to them because it is not available to others - it is quite an 'all or nothing' approach (Allhof and Lin). However, many individuals would not see this as particularly 'moral' because while we are practicing equality, this means that certain individuals will be left unnecessarily vulnerable to disease. Additionally, by definition, acts of racism, sexism, and other forms of discrimination were considered perfectly normal and 'moral' in the same societies that they would be definitively defined as immoral in the present day (Allhof and Lin). This highlights the fact that it is unwise to generally state that human enhancements are all ethical or unethical, moral or immoral. As aforementioned, a great amount of critical thinking and nuance is required to analyze each type of human enhancement and the context of the situation prior to making such decisions.

For the remainder of this chapter, the primary focus will be on the ethics of human enhancement, as they are more clearly defined and less personally normative than morals. However, in certain, if not many, instances ethics and morals do go hand in hand and as such morality will certainly be discussed in tandem with ethics.

Autonomy and Freedom

Freedom is a complex topic when discussed in nearly all contexts, a core pillar of democracy and commonly used in arguments made by proponents of human enhancements. This is specifically for enhancements that do not affect others, so those are chosen by someone for only themself and will have no tangible impact on anyone else. In this case, the argument is typically that in a democratic "free" society, individuals should thus be free to modify, change, or enhance themselves however they wish (Allhof and Lin). This stance is in some cases considered perfectly ethical and moral and is seen in different contexts of human enhancements, such as for example body modifications and plastic surgery. Of course, this goes hand in hand with one of the core values of bioethics and medical ethics, which is personal autonomy (Varelius). This describes leaving the power of choice regarding an individual's health and body to only themselves, given they are mentally capable of making that decision. Essentially, to be autonomous is to have freedom to make choices about your body and your healthcare without restrictions (so long as they do not harm others), and in medical practice this can look like choosing alternative treatments, or even denying treatment altogether (Varelius). However, there are certain contexts where fighting for freedom and autonomy actually results in being against human enhancement, and a very clear instance of this is with enhancement interventions taken for fetuses and children at conception, during pregnancy, and very early in life (Juengst and Moseley). The argument here is for the child, as by permanently 'enhancing' them genetically or biologically without their consent (which would be impossible to obtain until the child is of a certain determined age), you are then denying them their freedom and autonomy.

It is also important to note though when discussing freedom, that while it is a cherished aspect of democracy, it is also often limited for a variety of purposes. For example, individuals are not free to drive however they like, as in driving in the wrong lanes or ignoring traffic signals. Additionally, a popular example is freedom of speech, as while individuals are indeed free to generally say what they like, that does not exempt them from being prosecuted for libel, defamation, slander, or hate speech. In these cases, most members of a society can agree that certain freedoms should be limited to a certain degree for safety and protection of human rights of everyone in the society (Lin and Alhoff). Similarly, this poses the idea that while yes, individuals should theoretically be free to do what they like to their bodies as long as they are not directly impacting others, it is not always so clear cut. Certain enhancements may not directly impact others but definitely have the potential to negatively impact or disadvantage others or society as a whole, even if it is not intended.

Equity and Fairness

Understandably, fairness is a significant ethical issue when it comes to human enhancements. This is not only because of the argument that everyone should have equal access to the advantage that is enhancement, but also because it means that whoever is not 'enhanced' is then at a distinct disadvantage (Alhoff and Lin). Fairness is not a new issue within healthcare either, with private systems often criticized as being inherently unjust, but also with public healthcare systems often already stretched too thin with current demand. Thus, it would not be realistic to expect human enhancements to be covered by public systems, which further decreases accessibility and inequity (Miah). Additionally, the disadvantages for those who are not enhanced, depending on what the enhancement is, may be significantly detrimental to their livelihood and quality of life (Allhoff et al.). Imagine an individual enhanced to be a super-athlete, able to perform at a higher level than would ordinarily be possible. In this case, no other athlete would possibly have a chance at competing, thus limiting their success beyond their control.

However, it is important to distinguish between fairness and equality. As previously mentioned, an egalitarian approach to human enhancement ethics is often seen as impractical, as it is hardly ever a case where a service or intervention is only available when every single individual on Earth can access it. In fact, under most economic theories fairness does not imply equality at all - and gaps between economic classes are often viewed as desirable (Allhoff et al.). This is largely to establish incentives for innovation and upward mobility, and to be able to fill vacancies in the workforce where there are a wide range of tasks to be fulfilled (Alhoff et al.).

Human Dignity and Identity

Interestingly, in ethical debates surrounding the topic of human enhancement, human dignity is often seen used as an argument both for and against the issue. This brings forth the problem of "dignity talk", which is where two sides of an ethical debate, where human dignity is used as the 'argument ending trump card' (Kirchhoffer). This leads to a redundancy in debates, as if both sides use the same argument as their final play, there is ultimately no 'winner' of the debate.

However, one difficulty many face when using human dignity as an argument for either side of the debate is the fact that human dignity is a highly abstract concept that is difficult to define. Nonetheless, it is necessary to have a working definition for the purposes of discussing the concept and its role in the ethics of human enhancement, so the following by Lee and George will be used: "The dignity of a person is that whereby a person excels other beings, especially other animals, and merits respect and consideration from other persons" (Lee and George).

The clear argument from the conservative side -- as in the side opposing human enhancement -- is that human dignity is violated by enhancements. Some questions that are posed by that side include whether those who desire enhancements are inherently ungrateful for what they have and who they are, and further enable this unending cycle of dissatisfaction with one's life? (Alhoff et al.). Additional arguments include the fact that human enhancement would hinder character or moral development, with the belief that if one does not struggle and work hard for an achievement, it is ultimately meaningless (Lin and Allhoff). Another way of looking at this argument is that enhancement may alter human nature on a fundamental level, such that those who are enhanced lose their human nature or have a different form of it, ultimately impacting their human dignity (Giubilini and Sanyal). That is, if dignity comes from an individual having abilities and talents that they were either gifted' with from birth or alternatively worked hard for, then that dignity would be lost if those same abilities were a result of 'unnatural' enhancements (Giubilini and Sanyal, "The Ethics of Human Enhancement…").

On the other hand, the liberal side -- as in the side for human enhancement -- often argues that it is in fact part of human nature to constantly innovate and make alterations, and that this does not equate to a loss of dignity ((Giubilini and Sanyal, "Challenging Human Enhancement"). A modern example can be as simple as the invention of the internet and the technological revolution that followed, as this has certainly changed the way we behave fundamentally as humans and certainly made many tasks much easier. Many would argue that humans on the whole are more connected, more informed, and happier post-internet, so it may be argued that technologies related to human enhancement may actually lead to happier, more fulfilled lives (Lin and Allhoff). Even in the case of human dignity relying on the talents, abilities, and capacities of humans that distinguish them as human, it can be argued that enhancement would only serve to improve these abilities and the overall potential of humans. As a result, human enhancement in fact supports humans to thrive and therefore furthers human dignity (Kirchhoffer). One really good example of a real-life application of this argument is related to the debate surrounding life-extension technology, which is a subset of human enhancement. Proponents of life-extension and anti-aging technologies argue that aging brings forth a deterioration in "basic human characteristics" such as independence, identity, and social integration (Kirchhoffer). As such, if one must exhibit these characteristics to have human dignity, then to preserve human dignity is to prevent or delay the aging process for as long as possible (Baltes).

Ethical Case Study

In previous parts of this chapter, I provided numerous brief real-life examples that helped showcase the ethical dilemmas relating to each core argument used to either fight for or against human enhancement. Examples included enhancing athletes to perform at levels beyond what is ordinary and administering life-extending enhancements to aging individuals. In this part of the chapter however an introduction of a specific case study will occur, along with a deep dive into the ethical considerations of the situation to provide a more 'hands-on' approach to the topic.

The example 'case' that will be discussed is that of cognitive enhancement. This can be defined as "the amplification or extension of core capacities of the mind through improvement or augmentation of internal or external information processing systems" (Bostrom and Samberg). While cognitive enhancements have long been part of daily life in the form of schooling and education, mental training, and certain drugs such as nicotine and caffeine, one of the most controversial types of cognitive enhancements is pharmacological cognitive enhancement (PCE) (Maslen et al.). An example of a PCE is modafinil, a drug that was originally developed to treat narcolepsy, and has actually been shown to improve cognitive function in sleep-deprived and non-sleep-deprived healthy adults (Wesensten et al.; Turner et al.).

Discussing this case through the lens of human dignity and authenticity and the ethical issues in this regard, the argument against PCEs is quite clear (Maslen et al.). PCEs alter cognitive function, and it is an easy connection to make that if one alter's one's mind, then that person is no longer who they once used to be (Maslen et al.). This can be taken either in the direction of the individual simply no longer being themselves, or could go as far as saying this is no longer adherent to 'human nature' and therefore could that individual even be considered 'human' after their enhancement (DeGrazia).

However, a counter point of this lies in the ethical argument of autonomy and freedom, as individuals have long held the personal freedom and autonomy to pursue improving and bettering themselves cognitively. Is an individual who chooses to undergo cognitive behavioral therapy to overcome a phobia no longer the same person they were prior to therapy? (Butler et al.). Could we claim that they are no longer 'human' because they chose to enhance themselves in this way? In this way, one can argue that PCEs would in fact be promoting authenticity because individuals are autonomously choosing to enhance themselves in this way, and that one's abilities and skills are dynamic and change throughout their life (Maslen et al.).

Something to keep in mind with the autonomy argument is that it can be applied to support the opposing side of the debate as well. On the societal level, one must consider the ethical implications of the widespread use of PCEs. If humans were shown to perform at higher levels at cognitive tasks after taking PCEs, then would there be pressure from employers, educators, and society as a whole to use them?(Maslen et al.). In this case, this would be a breach of personal autonomy and an individual's choice to use or not use PCEs may no longer truly be their own, especially if instead of simply putting pressure on them to take it, an individual's employer may even begin to require it. This perspective and very real possibility certainly complicates the ethical debate and brings forth new considerations of what it truly means to have autonomy and how we can ensure one's choices are truly their own and unaffected by pressure, coercion, or the lure of social desirability.

Conclusion

Throughout this chapter, it was seen time and time again how nuanced and complex the ethical and moral considerations for human enhancements are. For each realm of ethics there are a host of perspectives to consider, and for each example of human enhancement there is an abundance of ethical realms to consider. While it is impossible to cover every ethical perspective and consideration relating to this topic in a single chapter, a brief overview of the most common and applicable considerations was provided. This included issues of freedom and autonomy, fairness and equity, and human dignity and identity. It is crucial to remember that there are arguments to be made using these concepts for both sides of the debate, those who whole-heartedly support human enhancements and those who vehemently condemn it. As such, it is incredibly difficult to determine whether or not, on the whole, human enhancements are ethical. Instead, the best practice would be to take each situation on a case-by-case basis, assess how the different ethical considerations would apply, and make a context-dependent decision.

References

Allhoff, Fritz, Patrick Lin, James Moor, et al. "Ethics of Human Enhancement: 25 Questions & Answers." Studies in Ethics, Law, and Technology, vol. 4, no. 1, Feb. 2010, p. 20121004. DOI.org (Crossref), https://doi.org/10.2202/1941-6008.1110.

Allhoff, Fritz, Patrick Lin, and Jesse Steinberg. "Ethics of Human Enhancement: An Executive Summary." Science and Engineering Ethics, vol. 17, no. 2, June 2011, pp. 201–12. DOI.org (Crossref), https://doi.org/10.1007/s11948-009-9191-9.

Baltes, Paul B. "Extending longevity: Dignity gain-or dignity drain?." MaxPlanckResearch 3 (2003): 15-19.

Bostrom, Nick, and Anders Sandberg. "Cognitive Enhancement: Methods, Ethics, Regulatory Challenges." Science and Engineering Ethics, vol. 15, no. 3, Sept. 2009, pp. 311–41. DOI.org (Crossref), https://doi.org/10.1007/s11948-009-9142-5.

Butler, A., et al. "The Empirical Status of Cognitive-Behavioral Therapy: A Review of Meta-Analyses." Clinical Psychology Review, vol. 26, no. 1, Jan. 2006, pp. 17–31. DOI.org (Crossref), https://doi.org/10.1016/j.cpr.2005.07.003.

DeGrazia, David. Human Identity and Bioethics. 1st ed., Cambridge University Press, 2005. DOI.org (Crossref), https://doi.org/10.1017/CBO9780511614484.

Giubilini, Alberto, and Sagar Sanyal. "Challenging Human Enhancement." The Ethics of Human Enhancement, edited by Steve Clarke et al., Oxford University Press, 2016, pp. 1–24. DOI.org (Crossref), https://doi.org/10.1093/acprof:oso/9780198754855.003.0001.

Giubilini, Alberto, and Sagar Sanyal. "The Ethics of Human Enhancement: Ethics Human Enhancement." Philosophy Compass, vol. 10, no. 4, Apr. 2015, pp. 233–43. DOI.org (Crossref), https://doi.org/10.1111/phc3.12208.

Grannan, Cydney. "What's the Difference Between Morality and Ethics?". Encyclopedia Britannica, 1 Sep. 2016, https://www.britannica.com/story/whats-the-difference-between-morality-and-ethics. Accessed 27 February 2023.

Juengst, Eric, and Daniel Moseley. "Human Enhancement." The Stanford Encyclopedia of Philosophy, edited by Edward N. Zalta, Summer 2019, Metaphysics Research Lab, Stanford University, 2019. Stanford Encyclopedia of Philosophy, https://plato.stanford.edu/archives/sum2019/entries/enhancement/.

Kirchhoffer, David G. "Human Dignity and Human Enhancement: A Multidimensional Approach: Human Dignity and Human Enhancement." Bioethics, vol. 31, no. 5, June 2017, pp. 375–83. DOI.org (Crossref), https://doi.org/10.1111/bioe.12343.

Lee, Patrick, and Robert P. George. "The Nature and Basis of Human Dignity." Human Dignity and Bioethics: Essays Commissioned by the President's Council on Bioethics, edited by Adam Schulman, vol. 21, no. 2, [President's Council on Bioethics, 2008, pp. 173–93.

Lin, Patrick, and Fritz Allhoff. "Untangling the Debate: The Ethics of Human Enhancement." NanoEthics, vol. 2, no. 3, Dec. 2008, pp. 251–64. DOI.org (Crossref), https://doi.org/10.1007/s11569-008-0046-7.

Maslen, Hannah, et al. "Pharmacological Cognitive Enhancementâ€"how Neuroscientific Research Could Advance Ethical Debate." Frontiers in Systems Neuroscience, vol. 8, June 2014. DOI.org (Crossref), https://doi.org/10.3389/fnsys.2014.00107.

Miah, Andy. "Ethics Issues Raised by Human Enhancement." In Values and Ethics for the 21st Century. BBVA, 2011.

Turner, Danielle C., et al. "Cognitive Enhancing Effects of Modafinil in Healthy Volunteers." Psychopharmacology, vol. 165, no. 3, Jan. 2003, pp. 260–69. DOI.org (Crossref), https://doi.org/10.1007/s00213-002-1250-8.

Varelius, Jukka. "The Value of Autonomy in Medical Ethics." Medicine, Health Care and Philosophy, vol. 9, no. 3, Dec. 2006, pp. 377–88. DOI.org (Crossref), https://doi.org/10.1007/s11019-006-9000-z.

Wesensten, Nancy J., et al. "Performance and Alertness Effects of Caffeine, Dextroamphetamine, and Modafinil during Sleep Deprivation." Journal of Sleep Research, vol. 14, no. 3, Sept. 2005, pp. 255–66. DOI. org (Crossref), https://doi.org/10.1111/j.1365-2869.2005.00468.x.

Chapter 4: The Social and Political Implications of Human Enhancement

By Hiya Goyal

Human enhancement, the idea of improving upon or transcending natural human limitations, has been a fascination of humankind for centuries. From the ancient usage of herbal and holistic remedies for both treating and preventing disease to the evolution of modern healthcare and pharmaceutical medicine, human enhancement has resulted in generally favorable health outcomes and consequently extended our life expectancies in unforeseeably profound ways. As innovations continue to reach new limits, humans have continued to enhance ways of life and extend life expectancies by enhancing knowledge and creating means of alleviating some of the burdens associated with the human condition. Activities such as physical fitness and innovations such as the invention of eyeglasses in order to address biological deficits have in many ways served as a cornerstone for the new era of human enhancement technology. In recent years, there have been many advances in fields such as biotechnology and nanotechnology that are transforming the way we live and work, ideally with the goal of easing responsibilities and optimizing one's efficiency (Rodríguez-Alcázar, 2017). However, these innovations are burdened by ethical and moral questions about the use of technology to enhance human abilities, and its potential impact on society as a whole. While policymakers and society are not concerned with every activity and intervention that might improve people's embodied lives, a cluster of debates in practical ethics that is commonly labeled as "the ethics of human enhancement" is coming to light with these technological advancements (Giubilini & Sanyal, 2015). Being that human enhancement technologies have the potential to transform the way we live, work, and relate to one another, these debates surround a myriad of necessary considerations to make before human enhancement can be implemented publicly and for commercial use. Additionally, this extends into posing a necessary conversation surrounding how the ability to biology and technologically advance the human condition can affect the existing social hierarchy and alter our social world (Giubilini & Sanyal, 2015). This includes concerns about the limits of legitimate healthcare and the new boundaries for an individual's right to both accept or reject their human alterations, as well as more general questions about distributive justice, science policy, and the public regulation of medical technologies. In this essay, we will be exploring the social and political implications of human enhancement, both for better or for worse, by exploring literature that discusses the existing limits to medical practices and then postulating how these limits would potentially be affected by advancements in human enhancement technology.

Before discussing the future of health and human advancements, it is important to first establish the scope of what healthcare looks like today and the popular perceptions of how medical treatments are perceived today along with their existing controversies. Medical treatment and human enhancement are both areas where science and technology have made significant strides, though the practices of both are constrained by legal and social barriers produced through the consideration of ethical implications of advancing individuals beyond the human condition. Understanding the existing limits to medical treatment and human enhancement is important for developing realistic expectations and identifying areas where further research and development are needed (Al-Rodhan, 2011). Treating and preventing disease involve medical interventions that aim to restore or maintain normal health by alleviating symptoms, curing illnesses, or preventing the onset of disease. These interventions are typically focused on addressing specific medical conditions or abnormalities that are deemed unhealthy or harmful. For example, treating cancer with chemotherapy or using a vaccine to prevent a viral infection are examples of medical interventions aimed at treating or preventing disease (Giubilini & Sanyal, 2015). On the other hand, human enhancement refers to interventions that aim to improve or augment human traits beyond what is considered normal or healthy (Al-Rodhan, 2011). These interventions are focused on optimizing performance or increasing capabilities, rather than addressing specific medical conditions. While there have been significant advances in areas such as prosthetics, cognitive enhancers, and genetic editing, there are still limits to what can be achieved, and there are risks and uncertainties associated with many of these human enhancement technologies.

Furthermore, the conversation surrounding ethics often permeates policymaking arenas who are concerned with the sociological and physiological risks associated with human enhancements, as they can raise questions about fairness, equity, and what it means to be human. Doctors are allowed to provide medical treatment and interventions that are intended to prevent, diagnose, and treat diseases or medical conditions. They are also permitted to offer treatments that alleviate symptoms or improve quality of life for patients. The medical boundaries that exist today are largely based on principles such as safety, efficacy, and medical ethics. Medical treatments and interventions are evaluated and regulated by government agencies and medical associations to ensure that they are safe, effective, and comply with ethical standards. When it comes to attempting to improve the human condition through enhancements, the medical boundaries become more complex (Al-Rodhan, 2011). While doctors may be permitted to provide certain enhancements that address specific medical conditions, they may be prohibited from providing enhancements that are intended to improve human traits beyond what is considered normal or necessary. This is because such interventions may not be considered safe or effective, or may raise ethical concerns related to fairness, access, and social cohesion. In addition, there are concerns that attempting to enhance human traits may result in unintended consequences, such as unintended side effects, unforeseen risks, or unforeseeable changes in social dynamics (Coenen et al., 2011). As a result, the medical boundaries for human enhancements should be more stringent than those for medical treatment and interventions, involving additional ethical considerations beyond safety and efficacy as well as serving as a platform for mandating more robust social and political implications.

As mentioned, the environment of standing medical practices is heavily regulated by political decisions and the de facto rules of society that view the risks of human enhancement as more potent than the benefits. However, it must be acknowledged that there are socially and politically driven reasons as to why human enhancement may be a viable avenue for bettering humanity as a whole. When it is possible for science to provide a solution for the imperfections associated with the human condition, it is possible to also fix some of the mistakes that have been made by humanity in the past (Howell, 2015). Human enhancement can also be used to promote personal growth and fulfillment, such as by improving self-esteem, emotional resilience, or creativity, which may be particularly relevant for people who feel limited by their current abilities or who wish to explore new opportunities (Dijkstra & Schuijff, 2016). This enhanced ability to pursue the pinnacles of what humanity has to offer can allow the human mind to assume the necessary skill set in order to become the perfect political figure; one which has the ability to rationalize utilitarian decisions and isolate personal sentiments from the decision-making process. Additionally, politicians who are concerned with national security may support human enhancement because they believe that it will give their country a competitive advantage in the areas of science, technology, and business (Dijkstra & Schuijff, 2016). Once again, by enhancing the human capacity for rationality and omitting some of the biological deficits that may arise as a result of nature or nurture, human enhancement could lead to economic growth and international prestige as well as refine a population and empower them against corruption, poverty and other social conditions (Dijkstra & Schuijff, 2016). The libertarian argument for "political minimalism" suggests that people should be able to make choices about their own bodies and minds, including whether or not to pursue enhancements, thus implying that the role of government should be completely removed from the choices that a person makes about their body (Rodríguez-Alcázar, 2017). Instead, it should become the onus of medical professionals to take steps to ensure biomedical safety protocols are followed when it comes to human enhancement and that the venture for its commercialization be put forth because in scientific terms, human enhancement can warrant more benefits than loss (Rodríguez-Alcázar, 2017). Medical professionals, more so than politicians, have the technical skills to consider establishing guidelines for patient safety and ethical behavior as well as create the policy necessary for ensuring informed patient consent, monitoring, reasonable expectations of outcome, and the possibility of harm or failure of treatments. Additionally, they can emphasize the need for transparency and communication with the patient, and make sure they are aware of the unknown risks of any treatments.

In addition to the political reasons for human enhancement, studies have shown that there is potent social reasoning that may suggest that human enhancement may overall be a benefit to humanity and the society that it has created. Human enhancement advocates may argue that access to enhancement technologies could help level the playing field for people with disabilities, or that cognitive enhancements could help address educational or income disparities (Hughes, 2010). For instance, intergenerational trauma is known to be genetically embedded in many marginalized and ostracized communities such as the Indigenous community and has contributed to social deficits within the community through facets such as drug abuse and deaths of despair. Human enhancement could be used to treat the intergenerational trauma of Indigenous communities by biologically treating their social endemics that policy cannot seem to address effectively. Instances of intergenerational drug abuse, depression and alcoholism can be rectified using gene therapy and ensure that subsequent generations of Indigenous individuals are not subject to the biological traumas that are associated with historical injustices against their community (Hughes, 2010). Additionally, human enhancement can treat the biological deficits that are associated with the human condition to greater extents than it is already doing so (Howell, 2015). Enhancements such as prosthetics, assistive devices, or gene therapies could help people with disabilities to have greater independence and a better quality of life. Rather than fund for alleviating the conditions of disability with technology such as prosthetics, with the correct genetic programming, it could be used to cure disabilities altogether and benefit individuals in overcoming some of the limitations of their condition.

However, there are also a number of political reasons against human enhancement as innovations in science and technology are often responded to with ethical discourse that prompts a contrarian view to those who advocate for the technology (Lin et al., 2014).. First, many people are concerned about the potential misuse of human enhancement. If people are able to enhance their physical and mental abilities, it could lead to an unequal society where those who can afford enhancements have an unfair advantage over those who cannot. This could lead to a situation where those who have access to enhancements have greater power and influence over those who do not. Second, there is the potential for ethical issues to arise from the use of human enhancement (Howell, 2015). For instance, if certain individuals were able to enhance their physical or mental abilities beyond the capabilities of the average person, it could create an ethical dilemma. Should these individuals be allowed to use their enhanced abilities to gain an unfair advantage over others? This could lead to a situation where those with enhanced abilities have access to opportunities that are otherwise not available to the average person. Human enhancement is the use of technology to improve physical and mental functions beyond the current capabilities of the human body. It is a rapidly growing field of research and development, with potential applications ranging from medical treatments to military applications (Lin et al., 2014). There is also the potential for human enhancement to have unintended consequences. If people are able to enhance their physical or mental abilities, it could lead to a situation where some people become too politically powerful and influential. This could lead to a situation where those with enhanced abilities have too much control over the lives of others (Lin et al., 2014). Finally, there is the potential for human enhancement to be used as a form of social control. If certain individuals or groups are able to enhance their physical or mental abilities beyond the capabilities of the average person, it could lead to a situation where those in power are able to control the lives of those without enhanced abilities. This could lead to a situation where those in power are able to manipulate and control the lives of those without enhanced abilities.

Human enhancement is a controversial topic that has been debated for many years. While some people believe that human enhancement technologies can help improve the quality of life, others believe that there are social reasons against human enhancement that make it a bad idea. The first reason against human enhancement is the potential for inequality. Human enhancement technologies can be expensive and out of reach for many people, resulting in a two-tiered society where those with access to the technology are more successful than those without which can in turn lead to deepening divides between the "haves" and "have nots", creating an even more unequal society (Trothen, 2017). Additionally, as is the case for many medical treatments that involve drugs or have some sort of immediate benefit, human enhancement is the potential for abuse. Human enhancement technologies, if used incorrectly, could lead to the exploitation of vulnerable people (Trothen, 2017). For instance, if someone has access to technology that can improve their physical or mental abilities, they could use it to gain an unfair advantage over others and repeatedly use their wealth in order to create enhancements that risk their life or compromise the health of their natural body (Hughes, 2010). What is more is that the chronic biological and health implications of the enhancement treatments may not be able to evident until years after their commercialization, during which policymakers may be at an impasse for reversing the effects of. In addition to the chronic biological changes that may be rendered unmitigated, human enhancement has the potential for large-scale sociological plights involving humanity losing their sense of identity and what it means to be human (Giubilini & Sanyal, 2015). Human enhancement technologies could lead to the homogenization of the population and perpetuate standards of appearance and ability that ostracize diversity because as people become more and more similar due to the use of these technologies, it could lead to a loss of individuality and unique characteristics. Finally, the fourth reason against human enhancement is the potential for unforeseen consequences. While human enhancement technologies may seem beneficial in the short-term, they could lead to unforeseen consequences in the long-term. For example, if people use these technologies to become "better" than everyone else, it could lead to a situation where those with access to the technology become dominant over those without it.

While human enhancement technologies may seem like a good idea in the short-term, there are a number of social reasons against human enhancement that should be considered before its commercialization. It is important to weigh the pros and cons of human enhancement technologies before implementing them. Human enhancement can be dehumanizing because it involves changing the human body in some way to make it stronger, smarter, or more capable than it naturally would be. This can lead to a devaluing of human life, as people may feel that they can no longer compete with those who have been enhanced. Additionally, people may be treated differently if they are enhanced, as some may view them as no longer being "fully" human. This could lead to discrimination, alienation, and a lack of respect for those who have been enhanced. However, human enhancement can also widen the bridge between the wealthy and the poor. Those with limited resources may be able to access enhancements that would otherwise be out of their reach and this could help to level the playing field and give those from disadvantaged backgrounds a chance to compete on a more equal footing with those from more privileged backgrounds.

Despite continued advances in technology and the potential for human enhancement, there remain major gaps in knowledge about how to use these advancements in a safe and ethical way. Researchers are faced with the challenge of ensuring that advanced medical treatments, computers, nanotechnology, and artificial intelligence all contribute to improved quality of life without manipulating human values, freedom, or dignity (Juengst & Moseley, 2015). Additionally, effective and responsible governance of these technologies raises long term ethical questions, such as how to manage the potential harms of enhanced individuals or how to protect their privacy. To ensure desirable and equitable outcomes, there must be global cooperation to ensure the protection of human identity, safety, and autonomy. As the implications of human enhancement remain largely unknown and the potential benefits of technological advancement are increasingly explored, gaps in knowledge about ethical and authentically useful human enhancement will continue to evolve (Dijkstra & Schuijff, 2016). These potential technologies, such as robotic prosthetics and gene editing, have the potential to create large disparities between those who can and cannot afford them. This could result in greater inequalities between the rich and the poor, and further losses of freedoms due to unequal access to certain technologies. In addition, those with greater access to advancements may be more likely to acquire further advantages across life domains, such as job opportunities, education, and social capital. This technology may therefore lead to a more stratified society in which privileged individuals have increased abilities and access, while those without the financial means to take advantage of such advancements are left at a disadvantage. To ensure that the commercialization of human enhancement does not further exacerbate the wealth gap, there must be policies in place to ensure access to technological advancements is equitable and that everyone has the same opportunity to benefit from them. This could include providing subsidies for those unable to afford treatments, or engaging in public-private partnerships in order to make treatments more accessible. Additionally, it is important to ensure that enforcement of regulations does not further single out certain populations or communities. Lastly, it is essential to make sure that any decisions regarding access to these treatments or technologies are based on need, not ability to pay. By engaging in these approaches and making sure technological advancements are made available to all, the effects of human enhancement on the wealth gap can be minimized.

References

Al-Rodhan, N. R. (2011). The politics of emerging strategic technologies: Implications for geopolitics, human enhancement and human destiny. Springer.

Coenen, C., Schuijff, M., & Smits, M. (2011). The politics of human enhancement and the European Union. Enhancing human capacities, 521-535.

Dijkstra, A. M., & Schuijff, M. (2016). Public opinions about human enhancement can enhance the expert-only debate: A review study. Public Understanding of Science, 25(5), 588-602.

Giubilini, A., & Sanyal, S. (2015). The ethics of human enhancement. Philosophy Compass, 10(4), 233-243.

Howell, A. (2015). Resilience, war, and austerity: The ethics of military human enhancement and the politics of data. Security Dialogue, 46(1), 15-31.

Hughes, J. (2010). TechnoProgressive biopolitics and human enhancement.

Juengst, E., & Moseley, D. (2015). Human enhancement.
Lin, P., Mehlman, M., Abney, K., & Galliott, J. (2014). Super soldiers (Part 1): What is military human enhancement?. In Global issues and ethical considerations in human enhancement technologies (pp. 119-138). IGI Global.

Rodríguez-Alcázar, J. (2017). Political Minimalism and Social Debates: The Case of Human-Enhancement Technologies. Journal of Bioethical Inquiry, 14(3), 347-357.

Trothen, T. J., & Mercer, C. (Eds.). (2017). Religion and human enhancement: death, values, and morality. Springer.

Chapter 5: Philosophical Perspectives on Human Enhancement

By Hannah Kuipers

Humans are constantly looking to improve their lives, which in turn improves society as a whole. One of the many ways we achieve this is through bioengineering, which creates new innovations and technologies that have the potential to better humanity. However, with this advancement comes the debate of ethics, social justice, the impact on human relationships, and so on. In this chapter we will explore different philosophical perspectives that help us understand this inquiry, and explain how bioengineering can have both positive and negative effects on our future. First, we begin by examining the utilitarian and deontological perspective by looking at ethics in medicine and other technological advancements. The ideology of utilitarianism states that the ethical thing to do is the action that brings the most happiness to the greatest number of people. In terms of human advancement, utilitarians believe that if it achieves pleasure and well-being for society as a whole, it should be pursued. However, there are cases where this ideology would argue against advancement through bioengineering, for example, if a new innovation would harm more people than it would help. As opposed to utilitarianism, deontology focuses more on our moral obligations, rather than just pleasure. Deontologists believe there is a clear right and wrong, and that there are rules about ethics that apply to all individuals, regardless of their culture or other circumstances. In terms of human advancement, this philosophical perspective believes that people should follow ethical rules, even if the consequences are uncertain or unpleasant. Next, we look at eudaimonism and virtue ethics by exploring the idea of transhumanism. Eudemonism, which comes from the Greek word eudaimonia, is commonly translated as happiness, yet it means much more than that. It aims to describe the highest human good for one's own sake. In relation to human advancement, it emphasizes individual happiness over the well-being of society as a whole. Rather than focusing on feelings such as satisfaction, virtue ethics will explain how human advancement through bioengineering relates to moral character development, and how this helps us cultivate good character traits. Lastly, this chapter will discuss ethics of care and existentialism. Existentialism is a theory that highlights the existence and experiences of individuals as a source of meaning in life. In this chapter existentialism is applied to bioengineering, and will explain how people look at and understand new technologies. Existentialism will also reflect on how it impacts personal identity. Looking on a larger scale, the ethics of care section will focus on feminism, and how feminist ideologies, such as relational feminism and the importance of relationships, play a role in human advancement.

Utilitarianism and Deontology

The theory of utilitarianism has been applied to many ethical dilemmas as it provides a moral framework for evaluating the benefits and costs of scientific and technological innovations. The goal of utilitarianism is to increase overall happiness and well-being, and advancements through bioengineering can play a key role in this. However, deontology provides an opposing view, as this belief advocates for using rules to define right from wrong throughout any process, not just when analyzing the rightness of an outcome. To distinguish the difference between the two, we will look at both utilitarian and deontological perspectives in ethics in medicine and other technological advancements.

Utilitarianism

Utilitarianism is a moral and ethical theory which states that the right action to take is the one that benefits the greatest number of people. To decide whether something is right or wrong, utilitarianists look at the outcome of a particular decision, rather than the choice itself. The first systematic account of utilitarianism was developed by Jeremy Bentham, alongside him were John Stuart Mill and Henry Sidgwick, who became what we know as classical utilitarians. They believed in the total view of population ethics, which regards one outcome better than the other if it brings a greater total well-being. They also accepted hedonism, a theory of well-being which states that welfare consists only as a balance of positive and negative experiences. However, more recent philosophers have argued that this view is too simplistic, as it is not human nature to feel only pleasure or only pain at a given moment. It is very likely that we may feel both at once, or rather not at all. The more recent view of utilitarianism is made up of four elements; consequentialism, welfarism, impartiality, and aggregationism. Consequentialism is arguably the main part of utilitarianism, it states that an action is judged solely by its consequences, as mentioned before. Consequentialism can be broken down even further, specifically into two parts, direct and indirect. Direct consequentialism, also known as act consequentialism, assesses the rightness of actions only. Indirect, or rule consequentialism, however, assess morality by the set of rules it follows to achieve the best utilitarian consequences. Welfarism, which is also known as well-being, is an important part of utilitarianism because it offers a more personal insight into the utilitarian idea of rightness. Well-being is defined as a state of comfortability and happiness, it takes into consideration things that bring us joy, not just things that are fundamentally important to us. It is crucial to understand this as utilitarians argue that the sum of individuals is a main determinant of the value of an outcome. Treating well-being as equally valuable as fundamental things in life is called impartiality, which is another aspect of utilitarianism. Impartiality also states that the well-being of individuals, regardless of nationality or gender, should be equal, and people should not privilege their well-being over others. Lastly, aggregationism, which looks at all other parts of utilitarianism in order to determine which choices are morally right to make. Simply put, it can add up certain factors, such as welfare, to decide which trade-offs are worth making. An example of this is the trolley problem, which asks participants if they would pull a lever to change the direction of a train to murder one person as opposed to letting it continue down its original path and kill ten people. Jeremy Bentham, a believer of classical utilitarianism, would agree with pulling the lever and sacrificing one person over a group of people, as this outcome would harm the least amount of people. However, many other philosophers would disagree.

Utilitarian and Deontological Perspectives on Human Advancement

Much of modern ethics deals with moral dilemmas arising in patient ethics such as their autonomy, confidentiality, consent principles, and so on. (Mandal, et al) The basic concept of utilitarianism is that the ends justify the means, and we should do anything to achieve the desired outcome. Specifically, rule utilitarianism says that morally right decisions are actions which comply with moral codes and rules to lead to better consequences. On the other hand, deontology views our obligations to be more important, in a way where the ends do not always justify the process. In contrast to the utilitarian concept, deontology is the ethics of duty where the morality of an action depends on the nature of the action.

In relation to human advancement, utilitarianism would support the development of new medical treatments and technologies if they led to an increase in overall health and well-being for the greatest number of people. This can include things such as the development of new medicines and drugs, or even genetically modified organisms used in farming or food quality. Alternatively, deontologists would consider the inherent rightness or wrongness of a particular bioengineering project, rather than simply its potential benefits or consequences. Because the basis of deontology places a strong emphasis on individual rights and autonomy, they could argue a new technological advancement is inherently wrong because it goes against the natural order of things. While these philosophical perspectives are quite different, they would both agree that if the development of any treatments or technologies through bioengineering would have negative consequences such as ethical concerns or creating a significant social inequality, it would be inherently wrong and not worth the process.

Eudaimonism and Virtue Ethics

Aristotle, an ancient Greek philosopher, is a name many of us have heard before. His array of theories are a part of our everyday lives, including the way we teach and understand, and the relationships we have with one another. He wrote two ethical treatises: Nicomachean

Ethics and Eudemian Ethics. In this chapter we will focus on Eudemian ethics, as this is what best encompasses our topic of Eudaimonism and virtue ethics, and how it relates to human advancement through bioengineering. Specifically, we will take a look at transhumanism, the belief that technology can be used to enhance human abilities beyond their biological limitations.

Eudaimonism

Eudaemonia is another word for a life "well-lived", and according to Aristotle, to achieve this humans must strive toward the highest good they could possibly be. In his mind, the highest good has three inherent qualities; it is desirable for itself, it is not desirable for the sake of some other good, and all other goods are desirable for its sake. (Kraut) The good of humans must also have something to do with being human, as this is what sets us apart from different species. As specified by Aristotle, this means that we have a potential to live a better life by guiding ourselves using logical thought and reasoning. However, Aristotle also acknowledges that we must be able to determine which good happiness consists of, as our understanding is limited when looking at simply one part of the whole. As a means to understand happiness, Aristotle addresses the question of ergon, or function, of human beings. In his book he states that ergon consists in the rational part of the soul, as activities which bring us happiness, such as seeing our friends and family, happens here.

Virtue ethics

Aristotle distinguishes two kinds of virtue, that which pertains to intellectual reasoning, such as virtues of the mind, and the part of our soul that cannot reason by itself, but is able to follow reason, such as ethical virtues. In his belief, we are all born with the potential to become ethically virtuous, but we must go through different stages of our lives to develop these habits. When we have developed proper habits we must acquire practical wisdom, which is no different from technical skills. Through these skills we learn how to do things such as intermediate between two extremes. The idea of knowing how to avoid excess and deficiency is a big part of Aristotle's overall ethical framework, which brings the topic of transhumanism into consideration.

Transhumanism

Transhumanism is the belief that technology can be used to enhance human abilities beyond their biological limitations. It includes advances in genetic, implantable, and wearable technologies, with artificial intelligence being one of the major examples. The questions we aim to answer in this chapter include; does transhumanism achieve eudaimonia? And, according to Aristotle's moral theories, is transhumanism ethically virtuous?

Eudaimonia, as discussed before, is a term to describe a "well-lived" life. Often it is translated to happiness and contentment. So, in order for transhumanism to achieve eudaimonia, it must bring satisfaction and well-being to humanity. There are already many advancements we have made through bioengineering that have greatly improved human life, for example, laser eye surgery. For individuals with glasses, ocular surgery has many benefits, including a decreased dependency on corrective eyewear, and long-lasting results with a quick recovery period. Another major example of transhumanist advancement is artificial limbs, which give people who have lost their own, through either disease or a traumatic experience, a chance to regain function and control of their body. This would substantially improve their quality of life. For instance, if an athlete lost a limb, they would lose their ability to play their sport. A prosthetic could give them the power to restart what they once loved doing, ultimately increasing their happiness. Things that expand our lifespan, such as vaccines and medicines, can also be considered human advancement, and significantly enhance humanity. Lastly, surgery, either for medical or cosmetic purposes, aims to achieve eudaimonia. Looking specifically at a rhinoplasty, it can improve well-being by giving the patient the ability to breathe better, but it can also improve happiness by giving them a heightened sense of self-esteem. Overall, these advancements create pleasure and satisfaction, so the answer to our first question; does transhumanism achieve eudaimonia? Is yes. However, we must also consider the negative aspects of transhuman advancement, which deals with the issues of virtue ethics.

The main point of Aristotle's theory of ethics is that virtue is in between two extremes. His principle belief is that a virtuous idea is a perfect balance between excess and deficiency. In terms of transhumanism, the extreme idea would be for technology to vastly enhance a person's intelligence; to tailor their appearance to what they desire; to lengthen their lifespan, perhaps to immortality; and to vastly reduce their vulnerability to harm. (McNamee, Edwards) The less radical side of this would be to simply use technology to enhance human characteristics such as beauty, lifespan, and resistance to disease. If all transhumanist concepts would fall directly in between these two categories, according to Aristotle, it would be virtuously ethical. However, since transhumanism is such a wide topic with many different objectives, the scope of this idea is simply too small.

Extreme transhumanism can create "possible future beings whose basic capacities so radically exceed those of present humans as to be no longer unambiguously human by our current standards". (Scholarly Community Encyclopedia) For many people, transforming the human species into a near-robotic being takes away the essence of what makes us truly human. (Kiebler). Inequality is also a concern many have about the rise of transhumanism. Technological advancements can be extremely expensive, and in some cases they are only available for people in the upper class. With these technologies being accessible to only a small percent of the population, the gap between the rich and poor could increase. On the other hand, biological engineering can also be used to aid in the increase of accessibility to human advancement technology. Lastly, drugs invented through bioengineering to aid in human advancement can and are being already misused. An example of this is the use of human growth hormone in non-clinical populations such as athletes. Not only is this unethical in terms of fair play, it is also extremely dangerous. So, the answer to our second question; is transhumanism ethically virtuous? Is, unfortunately, no.

Ethics of Care and Existentialism

Ethics of Care

Ethics of care is an ethical theory that is centered around interpersonal relationships and emotion-based decision making. This is very different from other ideologies discussed in this chapter because it takes on a more feminist approach. Popular ideologies, such as utilitarianism and deontology, are extremely male-centric due to the fact that decisions are made in unemotional ways. In most theories there is a defined set of rules that allow people to make rational and logical choices. In ethics of care, however, it is taken more on a case-by-case basis, and it applies emotional aspects such as care and compassion. It understands that conclusions are best made when taking into consideration the relationships we have with the people around us. Psychologist Carol Gilligan published a book called In a Different Voice, which argued that Sigmund Freud's theory of psychoanalysis and Lawrence Kohlberg's theory of moral development were biased and male-oriented. Freud's theory explains that our minds are composed of mental processes that we do not consciously think about, but that affect our feelings and behaviors. (Ackerman). Kohlberg's theory suggests that the reasoning behind the answers we give are a greater indication of moral development than the answer itself. (McLeod). The reason why Gilligan indicated these theories were male-centric is the fact female development is looked at inferiorly due to male development being the standard. (D'olimpio). It also goes back to the emotional versus unemotional debate, where she argues that making a choice based on emotions is not less ethical but simply a different approach. Educator Nel Noddings also defended care as a form of a moral relationship by explaining that care is ethically basic. It contrasts the ideas discussed earlier because it denies the idea that morality is strictly following universally logical rules.

Existentialism

Existentialism as a philosophical perspective emphasizes individual experiences and choices in shaping the meaning and purpose in life. One of the famous sayings in relation to existentialism is that existence precedes essence. (Aho) Basically, this states that we exist for ourselves, and we are always in the process of defining ourselves through the choices we make. A key concept existentialists believe in is freedom, where we create our identity and give meaning to our lives through our actions. While this is inescapable given the human condition, it is often accompanied by anxiety because we realize we alone are responsible for our choices and the consequences they carry. Lastly, the theory of existentialism argues that we cannot look down on the human condition from a detached, third-person perspective, as we are always taking some part in a self-interpreting event or activity.

Ethics of Care and Existentialism in Bioengineering

The perspective of ethics of care can be applied to bioengineering by considering the impact of new technologies on individuals and communities. Alongside this, existentialism comes in on a more individual level, which highlights the importance of individual responsibility and agency in shaping the development and application of new technologies. When looking at new technologies, such as genetic engineering, bioreactors, and tissue engineering, we would have to look at the accessibility and affordability of any new technology as well as the effectiveness of the treatment. The emotional component of ethics of care comes into play here because we realize that it is unethical to produce new technologies and medicines without creating a budget for helping low-income families. However, the existentialist perspective would argue for whatever is best in the meaning of life, personal identity, and our understanding of what it means to be human.

In conclusion, human advancement through bioengineering has sparked a variety of ethical dilemmas and philosophical debates. In this chapter we covered utilitarianism and deontology, which have offered different perspectives on morality. Utilitarianism focused on overall happiness while deontology focused on individual rights and principles. Eudaimonism, also known as the state of happiness or contentment, explained how people look at human advancement in relation to themselves, while virtue ethics described how it relates to our moral character development. Lastly, existentialism is a theory that emphasizes the existence and experiences of individuals as a source of meaning in life, and ethics of care highlights the importance of empathy and relationality in addressing the social and environmental impacts of bioengineering. These different philosophical perspectives help inform our understanding of human principles so we can apply them to advancement in all forms.

References

Marriam-Webster. "Trolley Problem." Words at Play, 11 Sept. 2020. https://www.merriam-webster.com/words-at-play/trolley-problem-moral-philosophy-ethics

MacAskill, William. "Elements and Types of Utilitarianism". Utilitarnism.net. University of Oxford. https://www.utilitarianism.net/types-of-utilitarianism/#the-four-elements-of-utilitarianism

Mandal, Jharna et al. "Utilitarian and Deontological Ethics in Medicine". National Center for Biotechnology Information, vol. 6(1): 5–7, pp. 724–729, 28 Jan, 2016. https://www.ncbi.nlm.nih.gov/pmc/articles/PMC4778182/

Phillips, Theresa. "Genetically Modified Organisms (GMOs): Transgenic Crops and Recombinant DNA Technology." Nature Education, 1(1):213, 2008. https://www.nature.com/scitable/topicpage/genetically-modified-organisms-gmos-transgenic-crops-and-732/

"Deontology". Ethics Unwrapped. University of Texas. https://ethicsunwrapped.utexas.edu/glossary/deontology

McNamee, M.J, Ewards, E.D. "Transhumanism, Medical Technology and Slippery Slopes". National Center for Biotechnology Information, 32(9): 513–518, Sep, 2006. https://www.ncbi.nlm.nih.gov/pmc/articles/

Kiebler, Jamie. "The Pros and Cons of Transhumanism". SoakMind. 14 Dec, 2019, Dec 13. https://soakmind.com/tech/the-pros-and-cons-of-transhumanism/

Serra, Miquel-Angel. "Human Enhancement and Functional Diversity: Ethical Concerns of Emerging Technologies and Transhumanism. Method Science Studies Journal, vol. 12, pg. 169-175, 2022. https://www.redalyc.org/journal/5117/51169287027/html/#:~:text=HE%20embraces%20the%20so%2Dcalled,longevity%2C%20and%20super%2Dwellbeing

Liu, Handwiki. "Outline of Transhumanism." Encyclopedia. 4 Nov, 2018. https://encyclopedia.pub/entry/32885#:~:text=Posthumans%20%E2%80%93%20in%20transhumanism%2C%20they%20are,being%20technologically%20evolved%20from%20humans.

"Aristotle's Ethics." Stanford Encyclopedia of Philosophy, 1 May, 2001. https://plato.stanford.edu/entries/aristotle-ethics/

Ackerman, Courtney E. "Psychoanalysis: A History of Freud's Psychoanalytic Theory". PositivePsychology. 7 May, 2018. https://positivepsychology.com/psychoanalysis/

McLeod, Saul. "Kohlberg's Theory of Moral Development." Simply Psychology, 2013. https://www.simplypsychology.org/kohlberg.html#:~:text=The%20three%20levels%20of%20moral,preconventional%2C%20conventional%2C%20and%20postconventional.&text=By%20using%20children's%20responses%20to,development20than%20the%20actual%20answer.

D'olimpio, Laura. "Ethics Explainer: Ethics of Care". The Ethics Centre, 16 May, 2019. https://ethics.org.au/ethics-explainer-ethics-of-care/

"Feminist Ethics." Stanford Encyclopedia of Philosophy, 27 May, 27. https://plato.stanford.edu/entries/feminism-ethics

Chapter 6: The Psychological and Emotional Implications of Human Enhancement

By Mya George

Human enhancement, when thought of from a cosmetic and aesthetic perspective, involves engaging in elective procedures, surgeries, and other processes to, from one's perspective, improve the visual appearance and potentially the functionality of one's features. In this chapter, we will consider the unique psychosocial motivations in this type of aesthetic enhancement including potential predispositions to seeking these types of procedures based on mental health and personality traits, the influence of social media, peer relationships, and other factors in addition to largely positive experiences in utilizing modern advancements in reconstructive surgeries.

Context of the Industry

Cosmetic plastic surgery is one of the most romanticized specializations in medicine with the ability of skilled surgeons and professionals to change one's appearance, reverse signs of physical aging, and blur the source of any insecurities. The surge in popularity of reality TV stars and other celebrities who have undergone cosmetic procedures and plastic surgeons themselves has garnered further attention for the field and common procedures. While plastic surgery includes procedures beyond aesthetics, including reconstructive, craniofacial, hand, and microsurgery, this chapter will investigate the psychological and social-emotional impacts of elective cosmetic procedures.

The world's largest association of facial plastic surgeons, the American Academy of Facial Plastic and Reconstructive Surgery (AAFPRS), in serving its members, found that there were an estimated 1.4 million procedures performed in 2021 (Anon n.d.). Among motivations expressed in the AAFPRS survey, a greater number of individuals were seeking both surgical (commonly rhinoplasty, facelifts, or blepharoplasty) and non-surgical procedures for aesthetic reasons. With the previous calendar year including the start of the COVID-19 pandemic, people were largely restricted to spending time at home and meeting for work and other purposes over video calling platforms which have people looking not only at others but also their own faces for hours on end (2021Statistics & Trends ReleasedDemand for Facial-Plastic Surgery Skyrocket).

Many professionals also saw patients that had been delaying their intended treatments or procedures, more eager to have them done after receiving their COVID vaccines and the end of lockdown measures (2021Statistics & Trends ReleasedDemand for Facial-Plastic Surgery Skyrocket).Some people were motivated to change their appearance after living through filtered video and social media realities for months during lockdown, while some also expressed wanting to feel refreshed and confident for the return to in-person events and work. While the motivations around pursuing cosmetic enhancement procedures is particularly interesting in the context of the COVID-19 pandemic, this unique case highlights common reasons that people seek these changes regardless of global health crises—one's insecurities, mental health, and comparison to others which is furthered by social media use. In the rest of this chapter, we will investigate all these influences on the decision to undergo cosmetic procedures, as well as one's satisfaction with the outcome, to develop an understanding of the psychological and social-emotional implications of human enhancement.

Pre-existing Mental Illnesses & Body Dysmorphia

Self-esteem, one's confidence and satisfaction with oneself, and self-concept, the mental image one has of themselves, both highly impact how one presents themselves in the world. From their personality and attitude to productivity, self-image can shape oneself in positive and negative ways. However, when one's self-perception is warped by societal standards or psychological conditions, one's relationship with their physical self can deteriorate. Furthermore, the face plays an important role in reflecting one's emotions when interacting with others. One's self image can be further impacted by the preferences and thoughts of others, and in extreme negative cases can contribute to anxiety, depression, other psychological disorders, and overall social avoidance that can often further one's mental decline (Shauly et al.).

One of the most common cosmetic procedures performed by plastic surgeons is rhinoplasty, commonly known as a nose job (Shauly et al.). Rhinoplasties are typically performed to change the shape of the nose to improve how one feels about their appearance or also to improve how one breathes through the nose. Rhinoplasty is often promoted to patients who are unhappy with the appearance of their nose to improve their mental, emotional, and functional well-being. However, in recent literature, psychosocial concerns that influence one to pursue cosmetic procedures have been considered in the context of post-surgical satisfaction and mental wellbeing. While common procedures like rhinoplasty can significantly improve the self-confidence of individuals with sound mental health, for people with severe anxiety, depression, or other psychological conditions, it is hypothesized that there is greater risk of negative outcomes. For clinically depressed patients, the satisfaction and excitement that comes from receiving cosmetic surgery is often not noted to last and some patients have been observed as undergoing painful procedures as a form of self punishment which in turn worsens their depression (Golshani et al.). One study surveying American adults found that adults with anxiety-related disorders, particularly generalized anxiety disorder and major depressive disorder, or body-dysmorphic disorder were more likely to seek aesthetic rhinoplasty. In finding this increased desire for enhancement procedures, the investigators strongly recommended that healthcare professionals assess their patient's mental health before proceeding with permanent changes and to integrate other treatment throughout their care pre- and post-operatively to "prevent successful outcomes secondary to mental illness" (Shauly et al.).

A primary challenge surrounding the care of individuals with select mental health conditions who are also seeking cosmetic plastic surgery is that the healthcare professionals they reach out to are rarely equipped to treat underlying complexities related to their mental illness–they are trained in surface level changes rather than what extends beyond physical appearance. People who want to change aspects of their physical appearance often consult with a dermatologist or plastic surgeon rather than general practitioners who would refer them to a psychiatrist or other relevant mental health providers. Thus, these patients are more likely to have their concerns addressed through direct action rather than therapy and addressing the original motivation for such drastic change. One study found that for individuals with body dysmorphic disorder, 46% were referred to dermatologists and 38% ultimately received dermatologic treatment rather than any psychiatric intervention (*Gender Differences in Body Dysmorphic Disorder:The Journal of Nervous and-Mental Disease*).Body dysmorphic disorder is commonly believed to be part of the obsessive-compulsive spectrum where one becomes distressed by self-perceived imagined or slight defects in their appearance. Since the condition can have a significant impact on psychosocial functioning, and has been associated with low quality of life and high risk of suicide, it is incredibly important that individuals with body dysmorphia receive treatment beyond cosmetic enhancements (Golshani et al.). When body dysmorphia is actually diagnosed by a mental health care professional, individuals can work on addressing compulsions related to their obsessive thoughts about their appearance through cognitive behavioral therapy or take medications commonly prescribed for anxiety-related disorders such as selective serotonin reuptake inhibitors (SSRIs), although there has been no randomized controlled trials comparing the efficacy of these two different lines of treatment (Golshani et al.). The primary issue regarding body dysmorphia at this point in time remains the underdiagnosis of the condition, based on individuals being hesitant to voice their concerns to healthcare professionals and the tendency for the type of professional to be reached out to being one whose job it is to correct physical attributes rather than address the whole of the individual.

Personality Types that Seek Cosmetic Surgery

The introduction and study of personality types in psychology has been the source of many debates within the discipline since the late 1960s, however, it remains important to acknowledge that both the characteristics of persons and situations have significant impacts on behavior (Pinon). Personality traits can be used to predict cognitions, emotions, and behaviors across different contexts and can be considered to judge human tendencies and improve upon faults in existing systems, notably in the medical system and beauty industry as it pertains to the psychological and emotional implications of human enhancement. In a study that collected data regarding participants' personality traits and their inclination towards elective cosmetic procedures and/or surgery, obsessiveness was the most common trait while being antisocial was the least common trait (Golshani et al.). Obsessive thinking lends itself towards making changes to one's appearance, as repeatedly considering facial features would likely motivate a person to take action and alter their appearance. One of the most interesting findings from the study was that considering a patient's personality traits played a role in their post-surgical satisfaction and consequently, personal confidence. Patients who were able to channel feelings of confidence and act happy, who were referred to as persons with a strong "faking it" ability in the study, felt the best once they had undergone a procedure to change their appearance. Ultimately, further considerations of personality characteristics and types can be useful in the reviewing process before drastic human enhancement–in consideration of how people will feel before, during, and after receiving this specific type of treatment.

In regards to specific personality types, narcissistic and histrionic personality disorders have both been commonly associated with elective cosmetic procedures and surgery (Golshani et al.). Both of these conditions are marked by the inherent need for attention from others, however, the motives differ between the two. Individuals with narcissistic personality disorder have an inflated sense of self and consequently believe that they deserve positive attention from themselves and others *(Signs of Narcissisticvs. Histrionic Personality DisorderCharlie Health)*. Individuals with histrionic personality disorder rely on outside validation to fuel their own self esteem, meaning that they struggle with being independent others. Previous studies have suggested that individuals with narcissistic tendencies and diagnosed narcissistic personality disorder are the most likely to seek botox injections and cosmetic rhinoplasty (Loron et al.). The higher frequency of individuals with narcissistic personality disorder seeking cosmetic enhancement aligns with their predicted goal of gaining more attention from others and meeting and even exceeding societal expectations of beauty. Individuals with histrionic personality disorder also have higher rates of cosmetic surgery, often with the goal to become more seductive or appealing to others as a means of fueling their self-esteem.

The Role of Social Media in Human Enhancement

Social media now plays a central role in the lives of many individuals. Whether you choose to actively engage with social media platforms or not, their influence is largely inescapable due to the sheer amount of individuals who use them on an everyday basis. Platforms where individuals share images and videos showing off their lifestyles, their belongings, and their physical appearance has the consequence of fueling negative social comparison and driving insecurities (Krause et al.). Pressures to look the same way as others or to present an even more polished and perfect appearance can drive individuals to alter their appearance using photo editing apps or software, to wear heavier makeup, to pose in certain ways to appear slimmer, and in serious cases, seek plastic surgery and/or cosmetic procedures.

One model created to describe three major influences on internalized understandings of beauty ideals is called the Tripartite Influence Model, which credits peers, parents, and the media with transmitting and reinforcing beauty preferences in individuals (Hardit and Hannum). Internalization of these ideals occurs when a person both accepts the external beauty standards and engages in behaviors to personally achieve that idea (Herring et al.). Media, including content from TV shows, magazines, the internet, and social media apps themselves, have a large role in shaping common societal views on body image. This includes ideas surrounding what body type is ideal and what clothing is fashionable and flattering, commonly expressed through imagery which can be edited and modified in print and publishing to achieve looks that are physically unattainable. The unique setup with social networking sites, specifically Instagram, where the focus is on images more than written text, is particularly interesting in the case of body image and the decision to receive cosmetic surgery. Instagram users are also encouraged to engage with posts by viewing, liking, and commenting on them–where looking conventionally attractive could have an important role in the amount of engagement one's posts receive (Herring et al.). Low-level appearance changes can be motivated by the desire for positive reactions on social networking sites, such as changing the style of clothes one wears, dying their hair, or changing how they apply their makeup (Herring et al.). Long-lasting changes such as cosmetic enhancements are likely to have more severe impacts on the individual, specifically in regard to their psychosocial functioning for better or worse (Herring et al.).

In addition to low self-esteem and negative thoughts towards one's physical appearance, the environment surrounding oneself can provide immense motivation to seek change through cosmetic procedures. Past experiences of bullying and teasing, which endures in online settings particularly through social media, have been well studied motivators for individuals seeking cosmetic surgery (Herring et al.). Additionally, hearing about individuals having undergone cosmetic procedures and especially voicing positive experiences with their decision is associated with individuals being more willing to go under the knife. Social media influencers and celebrities are often rumored to have undergone select procedures or openly talk about the artificial work that they have had done, making some users more likely to trust in the process (Herring et al.). Furthermore, many dermatologists and plastic surgeons themselves have found fame and a way to reach out to future clients through social media–making the influencer role take on whole new levels with highly educated individuals being able to utilize already impactful platforms. In a world where societal trends dictate how many individuals see and care for themselves, the role of social media marketing cosmetic procedures to people must be considered–especially when the most vulnerable users are likely young children and individuals with mental health conditions (as discussed previously in this chapter).

The Life-saving Potential of Reconstructive Surgeries

While this chapter has largely discussed cases of people wanting to or actually receiving cosmetic procedures to alter their appearance, we have yet to discuss cases where one has an altered appearance for a particular reason, whether it be cancer treatment or surviving an accident, and is able to receive reconstructive surgeries as survivors, not patients. This type of treatment is quite different from elective plastic surgery that seeks to perfect aspects of one's appearance. In the case of reconstructive surgeries, the goal is a return to normal - or a new normal - and to minimize the physical burden of past traumas. The motivation switches from looking in the mirror and wanting more to allowing oneself to process what happened to them and to foster healing in that way.

One of the most commonly thought of types of reconstructive cosmetic surgeries is breast reconstruction following a mastectomy, where the entirety of breast tissue is removed during cancer treatment or in high-risk circumstances where the individual is seeking preventative action. While there are less extensive surgical options to mastectomy, such as lumpectomies that remove only some tissue, there are circumstances where the cancer has progressed enough to make a mastectomy the safest option or where the individual themselves wishes for the more aggressive approach to decrease the likelihood of missing cancerous tissue (Foster and Wood). Regardless, after receiving a mastectomy, the patient has several options for rebuilding the shape of the breast if they wish. To mimic the natural appearance of the breast, saline or silicone-based implants can be used or autologous tissue can be used with or without artificial implants (Comprehensive Cancer Information - NCI). Implants typically require a two stage procedure where a tissue expander is inserted under the skin or chest muscle and slowly filled with saline. Once the chest tissue has been given enough time to heal, the actual implant is ready to be inserted (typically 2-6 months after one has undergone a mastectomy). For reconstruction using autologous tissue, tissue (called a flap) from elsewhere in the body that contains skin, fat, blood vessels, and sometimes muscle is used to give structure to the reconstructed breast. Flaps are usually from the abdomen or back but can also use tissue from the thigh or buttocks. In regards to the timing of starting reconstructive work, immediate reconstruction can start when the mastectomy is being performed, or it can be done once the incisions have healed and cancer therapy has been completed (Comprehensive Cancer Information - NCI). This second option is commonly called delayed reconstruction and can occur months or even years after the mastectomy (Comprehensive Cancer Information - NCI). While procedures to give a definite shape and size to the breast have improved over the years, no doubt helped by the market for elective breast augmentations or augmentation mammoplasty which has driven improvements and innovations, reconstruction of the nipple and areola has remained more challenging. The nipple is usually reconstructed 3 months post operation to achieve ideal placement (Silsby). Tissue can be used from a flap using the remaining nipple if it is large enough and the person has only had a single mastectomy, or using tissue from the labia (Silsby). To improve the realism of the reconstructed breast, nipple-areola tattooing has become popular following reconstructive surgery (Nipple Tattoos and Breast Cancer: Cost and How to Get a Tattoo). Experienced tattoo artists are able to create hyper-realistic nipples and areolas that appear 3D–offering cancer survivors the opportunity to closely match the visual appearance of their chest prior to their diagnosis and demanding treatment.

Understanding the emotions surrounding breast reconstruction is relevant not only to caring for patients, but for directing future research aims and advances in plastic surgery. Considering women's experiences in this context is crucial for ensuring that they have adequate support and the ability to heal from serious medical experiences. While many medical systems and local guidelines give specific instructions on the amount and type of information to share with cancer survivors undergoing mastectomy and reconstruction, the reality is that many still, in retrospect, have felt unprepared for what the surgeries and healing process were actually like. Individuals have reported not expecting the extent to which scarring changed the appearance and feel of their breasts, and that the differences between their expectations and reality directly influenced dissatisfaction and negatively impacted their mental health in the recovery stages (Herring et al.). Additionally, there is a need for healthcare professionals to place a greater emphasis on the psychosocial wellbeing of patient survivors in the reconstruction process, both before and after their operations. Considering individuals for their unique experiences and needs is necessary due to the inherently personal nature of these types of cosmetic procedures. This includes considering the content and mode in which information is provided to patients, and their preferences for care and how it is discussed (i.e. with family or other supports in the room or not). Ultimately, reconstructive surgery has the ability to change lives for the better, but the desires of the patient must be considered along with prioritizing patient comfort throughout the process, including consultation and the multiple stages of treatment.

Conclusion

The decision to seek and ultimately receive cosmetic enhancement through surgical and non-surgical procedures is not independent from outside influences. One's pre-existing mental health and overall health state informs other vulnerabilities that may influence their decision, in addition to specific attention-based personality traits. The decision to drastically change one's appearance might be based on a troubled sense of how one sees themselves and their body, or might be based on outside opinions or a desire to feel more desirable to others. Conversely, there can be cases where one does not have significant psychosocial impacts from receiving cosmetic procedures, or where surgeries may improve one's relationship with themself and their body. Regardless, appearance has the potential to have a significant impact on one's mental state and social relationships and consequently the cosmetic enhancement field needs to better consider the unique perspectives of their patients throughout the provision of care.

References

2021 Statistics & Trends Released - Demand for Facial Plastic Surgery Skyrocket. https://www.aafprs.org/Media/Press_Releases/2021%20 Survey%20Results.aspx. Accessed 11 Feb. 2023.

Comprehensive Cancer Information - NCI. https://www.cancer.gov/. Accessed 28 Feb. 2023.

Foster, R. S., and W. C. Wood. "Alternative Strategies in the Management of Primary Breast Cancer." Archives of Surgery, vol. 133, no. 11, Nov. 1998, pp. 1182–86, doi:10.1001/archsurg.133.11.1182.

Gender Differences in Body Dysmorphic Disorder : The Journal of Nervous and Mental Disease. https://journals.lww.com/jonmd/Abstract/1997/09000/Gender_Differences_in_Body_Dysmorphic_Disorder.6.aspx. Accessed 28 Feb. 2023.

Golshani, Sanobar, et al. "Personality and Psychological Aspects of Cosmetic Surgery." Aesthetic Plastic Surgery, vol. 40, no. 1, Feb. 2016, pp. 38–47, doi:10.1007/s00266-015-0592-7.

Hardit, Saroj K., and James W. Hannum. "Attachment, the Tripartite Influence Model, and the Development of Body Dissatisfaction." Body Image, vol. 9, no. 4, Sept. 2012, pp. 469–75, doi:10.1016/j.bodyim.2012.06.003.

Herring, Beth, et al. "Women's Initial Experiences of Their Appearance after Mastectomy and/or Breast Reconstruction: A Qualitative Study." Psycho-Oncology, vol. 28, no. 10, Oct. 2019, pp. 2076–82, doi:10.1002/pon.5196.

Krause, Hannes-Vincent, et al. "Unifying the Detrimental and Beneficial Effects of Social Network Site Use on Self-Esteem: A Systematic Literature Review." Media Psychology, Aug. 2019, pp. 1–38, doi:10.1080/15213269.2019.1656646.

Loron, Alireza Mohebbipour, et al. "Personality Disorders among Individuals Seeking Cosmetic Botulinum Toxin Type A (BoNTA) Injection, a Cross-Sectional Study." The Eurasian Journal of Medicine, vol. 50, no. 3, Oct. 2018, pp. 164–67, doi:10.5152/eurasianjmed.2018.17373.

Nipple Tattoos and Breast Cancer: Cost and How to Get a Tattoo. https://www.medicalnewstoday.com/articles/nipple-tattoo-breast-cancer. Accessed 15 Feb. 2023.

Pinon, Adelson. "Why Most Psychologists Should Assess and Report Personality." Frontiers in Psychology, vol. 10, Aug. 2019, p. 1982, doi:10.3389/fpsyg.2019.01982.

Shauly, Orr, et al. "Assessment of Wellbeing and Anxiety-Related Disorders in Those Seeking Rhinoplasty: A Crowdsourcing-Based Study." Plastic and Reconstructive Surgery. Global Open, vol. 8, no. 4, Apr. 2020, p. e2737, doi:10.1097/GOX.0000000000002737.

Chapter 7: The Legal and Regulatory Issues in Human Enhancement

By Paige Breedon

Introduction

The following chapter intends to outline the rationale behind the regulation of human enhancement, existing legal and regulatory frameworks and the major issues that arise when such regulations are set into motion.

Defining Human Enhancement

According to bioethics, human enhancement is defined as genetic, biomedical or pharmaceutical interventions to enhance human dispositions, capacities and well-being (Giubilini 1). Note that this definition does not require the individual being improved to be suffering from disease or have some underlying pathology in need of treatment (Giubilini 1). Therefore, human enhancement is not limited to life-saving medication or therapy. Instead, it can be applied to individuals who are disease free with the intention of giving them some advantage. Understandably, many ethical problems arise when such technology is used. Thus, various legal and regulatory frameworks are in place to attempt to control and manage advances in technology and the use of human enhancement.

Firstly, it is crucial to recognize the rationale for having legal and regulatory frameworks in place for human enhancement. For instance, from an ethical perspective, letting human enhancement proceed without restrictions could lead to a world where human rights, dignity, justice and equality are compromised (Giubilini 1).

For instance, in the case of genetic enhancement, two major issues arise given the outcome of such a procedure: discrimination against people who are different and superiority of transhumans (Raposo 189). If society progresses to the point where procedures such as genetic enhancement are the norm, those who abstain from such procedures or who have abnormalities or disabilities that cannot be fixed by an existing technology and thus remain in society will become vulnerable to being discriminated against (Raposo 189). This would ultimately lead to a lack of genetic diversity and grave stigmatism against unmodified members of society (Raposo 189). Also, for those that do elect to become modified and enhanced, and thus could be referred to as transhumans, there is fear that they could be seen as superior and eventually lead to natural humans being obsolete and perhaps dying off (Raposo 189). Some might even suggest that the rise of transhuman could affect the homo sapiens species similar to how homo sapiens drove Neanderthals to extinction (Raposo 189). Thus, there are immense social issues that could arise due to the employment and normalization of human enhancement as well as the sacrifice of humankind's dignity and integrity as a species.

Existing Legal and Regulatory Frameworks for Human Enhancement

Today, many countries have laws to control biomedical developments; such laws vary in severity from permitted for the sake of research to strictly prohibited with accompanying criminal charges (van Beers 1–36). For instance, Canada, Europe, Australia, and Brazil are countries that prohibit germline modification and countries like China, the USA and UK permit such technology; they are heavily regulated (van Beers 1–36). In general, many countries vary in how and to what extent that they regulate human enhancement.

Firstly, it is important to acknowledge if legal and regulatory frameworks for human enhancement vary based on the intention behind the enhancement (i.e. if it is for life-saving purposes or for elective cosmetic purposes). Some of the literature suggests that it is unclear whether legal and regulatory frameworks vary based on intention behind enhancement. Notably, an article by Raposo (189) acknowledges a common concern against genetic enhancement is that it will create some sort of "genetic apartheid". However, there is much to weaken such an argument against genetic enhancement because it furthers existing inequalities. For instance, Raposo (189) suggests that there is this underlying inequality based on genetic inheritance that always exists in society, i.e. some individuals are born with natural talents or characteristics that others are not due to winning the genetic lottery. Also, Raposos (189) proposes that the problem of a potential "genetic apartheid" could be mitigated if any new technologies or means of genetic enhancement were offered to everyone. However, the problem then becomes the affordability of such a suggestion, in most cases national health services or private health insurance packages do not currently pay for genetic enhancements (Raposo 189). There are also many inconsistencies in funding from such sources, as Raposo (189) suggests that abortion which is not a disease or abnormality is funded under most national health service providers but a mammary reconstruction after a breast excision from cancer treatment is not covered as it falls under cosmetic surgery (Raposo 189). Thus, for regulation of genetic enhancement to be viable, the funding framework of public and private health insurers may have to be modified if the way viability is achieved is through equal access among citizens.

Another area subject to legal authority is whether or not a technology or method of human enhancement can be patented and under what circumstances would this be permitted. For the European Patent Convention (EPO) with respect to human enhancement (i.e. Article 53(c)), strict rules exist that regulate patents issued and they vary depending on if the human enhancement is a therapy, for cosmetic purposes or for other intended purposes (Nordberg 19–28). In general, the EPC does not permit patents to be issued when there is danger that might interfere with the freedom of the medical profession to diagnose and provide therapy to patients as this could interfere with access to methods for such endeavors (Nordberg 19–28). More specifically, for interventions pertaining to treatment of humans or animals by surgery, therapy, or diagnosis, EPC will not permit patentability (Nordberg 19–28). However, for treatments that are non-functional, or for only aesthetic purposes, treatments may be patentable (Nordberg 19–28).

Also, a contrast in regulation that is made between prenatal genetic enhancement and postnatal genetic enhancement needs to be distinguished. Specifically, when postnatal rather than prenatal genetic enhancement is made there are putative obligations of the state to the actual persons instead of future persons thus making such investigation less speculative (Tamir, 576–610). A paper by Tamir (576–610) examined the parens patriae doctrine of postnatal genetic enhancement of minors in both protective and substitutive cases and discussed the role of the state in regulation of such enhancement. Just as with Nordberg (19–28)'s suggestion that cosmetic enhancement is patentable, Tamir (576–610) suggests any sort of cosmetic, elective surgery is also a serious concern and under lots of criticism for use on minors. These concerns and criticisms perhaps, suggest a greater need for state regulation for both protective and substitutive purposes. The article also suggests the parens patriae doctrine comes into play if postnatal genetic enhancement-specific harm were to occur. Harm includes mental harm, harm to the child's identity or human dignity, social harm, or causing over qualification (i.e. becoming superior in one's skills). The article makes note that liberal states where personal freedom and liberty is emphasized most, regulates and controls chives of postnatal genetic enhancement less even if suspicion of postnatal genetic enhancement-specific harm comes into play. Overall Tamir (576–610) emphasizes the limitations in the regulation of genetic enhancement for minors and how differences in pre or post-natal enhancement as well as societal values can influence such regulation.

Government or agencies can regulate the use of human enhancement technologies by controlling how public funding is attributed towards these procedures. For instance, genetic enhancement may not be funded under Medicare programs (Mehlman 515–27). Also, private health insurance plans do not fund cosmetic medicine, thus forcing clients to pay out of pocket. Paying out of pocket, may hopefully, dissuade many people from using such services. However, this means of regulation becomes problematic, as it may lead to the socioeconomic gap widening. This occurs as now only people who are wealthy enough or of high enough socioeconomic status can afford such services, leaving those less privileged or well off to be unable to obtain these services. The means of regulation by controlling funding may be not necessarily motivated solely because of limiting human enhancement use, but rather it may be financially motivated, as even if government was interested in paying for cosmetic surgery, it would cost them billions of dollars to afford such services for their population and thus is not a feasible option (Mehlman 515–27).

Problems With regulating Human Enhancement

Violating Human Rights - CRISPR Babies Case Study

Clustered Regularly Interspaced Short Palindromic Repeats (CRISPR) were first discovered in archaea and later bacteria by Francisco's Mojica in Spain ("Questions and, " 2023). It is an essential part of the bacterial immune system if infected by a virus serving as a way for the bacteria to store "memories" of past infections ("Questions and, " 2023). Then upon reinfection by the same virus, such "memories" can be located and utilized for protection. Then by 2013, Feng Zhang and colleagues published the first method to engineer CRISPR to edit the genome of mouse and human cells ("Questions and, " 2023). CRISPR technology works by allowing a researcher to design a sequence of RNA that matches a place on the genetic sequence where they want to insert DNA or inactivate a gene and pair that with a Cas9 protein capable of cutting the DNA ("Questions and, " 2023). Ultimately, this technology is very powerful as it allows a researcher to insert corrected versions of genes or inactivate genes entirely. It can be used to conduct experiments such that researchers can determine what genes are involved in what processes via the selective inactivation of different genes ("Questions and, " 2023). As such technology advances further, use such technology as interventions in the population to correct diseases with underlying genetic etiology; for instance, cystic fibrosis, caused by a three-nucleotide deletion, could be resolved by such technology. Although CRISPR technology offers much hope as an intervention to genetically caused diseases, it poses a significant threat to human dignity and equality if utilized inappropriately.

For instance, in November 2018, a biophysicist named He Jiankui announced that he had created two genetically modified babies using CRISPR-Cas9 technology to modify the DNA of their embryos before implantation for pregnancy (van Beers 1–36). Such modifications were aimed at making the babies resistant to infection from HIV; thus they were done in a preventative effort involving eight couples with male partners who were HIV positive (van Beers 1–36). This announcement led to a surprised and horrified reaction among the scientific community, as Jiankui's ignorance in the face of ethical and scientific conventions, fundamental rules of research on humans and norms of medical practice had been utterly disregarded (van Beers 1–36). Especially considering that the technology he was employing had not been cleared forwork on human subjects and was still in the experimental stages, needing to be further investigated to ensure safety and efficacy. The benefit-to-risk ratio was very unfavorable, and this case emphasizes the need for stringent regulation. Today Jiankui is a disgraced scientist but the question remains what kind of precedent did his actions take? Will there continue to be scientists and agencies going against the grain to provide technological advancements?

Balance Between Regulation and Innovation

As technologies advance, it can become harder to maintain such stringent legal and regulatory requirements over human enhancement. It can also take much work to balance legal and regulatory frameworks while promoting technological and medical innovations. For example, a paper by Schermer and colleagues (75–87) looking at the future of psychopharmacological enhancement concerning expectations and policies discussed whether governments should prohibit or promote enhancement research. Those on the side of innovation suggest that it is inspired by a utilitarian perspective, the idea that if a particular development benefits someone individually and in society as a whole, it ought to be pursued; those may go as far as to say that there is a moral duty to pursue such advancements (Schermer et al 75–87). However, Schermer and colleagues (75–87) acknowledge that this argument often fails to recognize the lack of sufficient empirical evidence about whether an intervention/treatment is beneficial as a lot can be unknown about the general effects and side effects.

They also suggest that it would be unjust to pursue more novel treatments and technologies over fundamentally life-saving treatments in certain situations. For example, in developing countries, prioritization should be given to life-saving treatments backed by sufficient empirical evidence as opposed to more elective human technologies that are less developed and seem less necessary in comparison (Schermer et al 75–87). Specifically, anti-malaria medication is a greater priority over cognitive enhancers (Schermer et al 75–87). However, Schermer and colleagues (75–87) advocate for freedom of research as an essential end and side with those advocating for increased freedom in research and working towards advances in pharmacology; however, they acknowledge the immense regulations required in order to apply such research and implement the use of psychopharmacological enhancement.

Notably, an article by van Beers (1–36) acknowledges that a growing number of people are fighting to lift bans on heritable genome editing for therapeutic purposes as soon as technology has been identified as safe to use on patients. Van Beers (1–36) even suggests that germline gene editing has become a race in that different countries or groups in societies are in an attempt to out-discover one another. Such a race can be seen in the age of do-it-yourself CRISPR kits being sold, thus allowing anyone from the comfort of their home to experiment with their DNA (van Beers, 1–36). These opportunities are vital for those who are part of a new biohacking movement inspired by bringing an end to the biotechnology sector's monopoly on genetic engineering (van Beers, 1–36). Van Beers (1–36) suggests that the start of the race towards further germline editing began in April 2015 when a Chinese stem cell researcher, Junjiu Huang and his colleagues published an article explaining their attempt at modifying human embryos through genetics. The prospect of such a race between science, industry and society, as well as the emerging movement of biohackers, poses a significant dilemma and challenge for those developing and maintaining legal and regulatory frameworks on human enhancement.

In Canada, under the Assisted Human Reproduction Act of 2004, editing the human genome carries a sentence of up to ten years in prison (Schaub). Thus, no altercations are to be made to the human genome under Canadian law if it can be herediable (i.e. to germ line cells) (Schaub). Although this caveat still permits individuals to edit the genome of skin or muscle cells for purposes of research, it limits any changes that could be genetically passed onto an offspring, an area of research that Canadian law is essentially condemning through their actions. More recently however, there has been a big push to allow for more freedom for researchers to explore gene function, embryonic development and reproduction (Schaub). Thus, yet another dilemma is posed to lawmakers as they must balance the demands of scientists and researchers with those that appease ethicists and human rights activists. A Stem Cell Network member Dr. Bartha Knoppers is someone who advocates for more freedom to allow for scientific discoveries, even going as far to argue that Canadian law violates the Universal Declaration of Human Rights set forth by the United Nations because it limits people's right to benefit from scientific discovery (Schaub). She makes arguments that other countries have more relaxed laws on gene editing, or prefer a regulatory approach that first ensures the safety, quality, and impact on human rights restrictions of new technologies before permitting their use in controlled contexts (Schaub.). Although it seems that Dr. Knoppers is putting much blame for lack of scientific discoveries on the Canadian government, it is also clear that safety of nuanced technology has to be critically evaluated and ensured before proceeding to be used in Canada (Schaub). She acknowledges technologies such as CRISPR, as they have great promise but require much more exploration and safety testing to be done prior to use on germ line cells (Schaub).

Issues From a National Defense/Military Perspective - Canada Case Study

A report titled "Biotechnology, Human Enhancement and Human Augmentation: A Way Ahead for Research and Policy," published in 2021 by Defence Research and Development Canada, offered the perspective of adopting human enhancement for the sake of national defense, i.e. offering services to enhance military personnel. Specifically, it acknowledged security and compliance considerations, including those affecting society and technology adoption's temporal and permanence considerations (Adlakha-Hutcheon et al). The paper acknowledges that from the defense perspective, novel enhancement technologies threaten human values of character and virtue in service and pose the risk of personnel or limitation of their role (Adlakha-Hutcheon et al). From the societal perspective, civilians might have concerns about members of society becoming superior through enhancement, which may lead to a gap in understanding and appreciation between civilians and military personnel (Adlakha-Hutcheon et al). Also, different legal and regulatory frameworks may apply depending on the invasiveness and advancement of the technology (Adlakha-Hutcheon et al). For instance, wearable technology or nutraceuticals may be appropriate given existing legal, ethical and political frameworks (Adlakha-Hutcheon et al). However, more invasive technologies, such as genetic editing, may require a new legal framework requiring immense efforts from lawyers, medical professionals, and bioethicists (Adlakha-Hutcheon et al). Also, concerns emerge regarding compliance among military personnel due to fear of benefit-to-risk ratio or personal reasons. Security threats such as biohacking must also be considered in an emergency; it would be highly problematic if other countries or agencies could sabotage entire defense populations (Adlakha-Hutcheon et al). Some of the proposed recommendations from the NATO Human Factors and Medicine Panel include standing up an operational bioethics panel, conducting further research on the development and use of technologies, and strengthening and further developing appropriate frameworks that consider ethical, legal, social, and environmental concerns (Adlakha-Hutcheon et al). Overall, using human enhancement in the context of military defense and national security poses unique challenges and demands regulation to ensure the safety and security of countries and societies.

Conclusion

Overall it is essential to acknowledge the many challenges faced by those who intend to regulate and legally restrict the use and development of human enhancement technologies. As discussed, such problems include balancing the conflicting demands for innovation in science and technology with preserving societal values and human dignity, growing demands among scientists and industry to advance humankind, instances of malpractice and ignorance towards humankind and regulating such technologies with the purpose of national security and defense. Compromising with innovators and ethicists requires further research. Collaboration should be conducted to ensure safety, efficiency, and ethics are appropriate and fight against the commercialization of not pathologically motivated human enhancement. However, such efforts may be met with great frustrations and resistance by various industries and agencies.

References

Adlakha-Hutcheon, Gitanjali., Richins, Matthew T., & Talor, Deborah E. Biotechnology,
Human Enhancement and Human Augmentation: A Way Ahead for Research and Policy. Defence Research and Development Canada, 2021. https://cradpdf.drdc-rddc.gc.ca/PDFS/unc386/p814643_A1b.pdf

Giubilini, Alberto., Sanyal, Sagar. Challenging human enhancement. In: The ethics of human
enhancement: understanding the debate; 2016. pp. 1.

Mehlman, Maxwell J. "Human Enhancement Uses of Biotechnology, Law, GeneticEnhancement, and the Regulation of Acquired Genetic Advantages." Encyclopedia ofEthical, Legal and Policy Issues in Biotechnology, vol. 1, 2000, pp. 515–27, https://doi.org/10.1002/0471250597.mur014.

Nordberg, Ana. "Patentability of Methods of Human Enhancement." Journal of IntellectualProperty Law & Practice, vol. 10, no. 1, 2015, pp. 19–28,https://doi.org/10.1093/jiplp/jpu203.

"Questions and Answers about CRISPR." Broad Institute, 18 Jan. 2023, www.broadinstitute.org/what-broad/areas-focus/project-spotlight/questions-and-answers   about-crispr.

Raposo, Vera Lãocia Carapeto. "The Better I Can Be: In Defence of Human Enhancement for a New Genetic Equality." Canadian Journal of Bioethics = Revue Canadienne de Bioéthique, vol. 5, no. 2, 2022, pp. 189–, https://doi.org/10.7202/1089801ar.

Schaub, Ben. "Human Gene Editing Could Change the World - What Are the Laws Governing Itin Canada?" CBCnews, CBC/ Radio Canada, www.cbc.ca/natureofthings/features/gene-editing-in-canada#:~:text=Editing%20human%20genes%20is%20restricted%20 in%20Canada&text=Under%20the%20Assisted%20Human%20 Reproduction,way%20that%20could%20be%20inherited.

Schermer, Maartje, et al. "The Future of Psychopharmacological Enhancements: Expectations
and Policies." Neuroethics, vol. 2, no. 2, 2009, pp. 75–87, https://doi.org/10.1007/s12152-009-9032-1.

Tamir, Sivan. "Postnatal Human Genetic Enhancement and the Parens Patriae Doctrine." Journal of Law and the Biosciences, vol. 3, no. 3, 2016, pp. 576–610, https://doi.org/10.1093/jlb/lsw039.

van Beers, Britta C. "Rewriting the Human Genome, Rewriting Human Rights Law?: Human Rights, Human Dignity, and Human Germline Modification in the CRISPR Era." Journal of Law and the Biosciences, vol. 7, no. 1, 2020, pp. 1–36, https://doi.org/10.1093/jlb/lsaa006.

Chapter 8: The Economic Implications of Human Enhancement Outline

By David Henneberg

Brief Overview of Human Enhancement

Human enhancement can be put into any of the following categories: bioelectronics, senses, brain interventions, bionics, exoskeleton, genetic modification and pharmaceuticals. Bioelectronics is the study of electronics intertwined with biology (human or otherwise). Senses refers to any technology being used to repair or enhance the human ability, whether that is hearing, seeing, touching, smelling, or tasting. Brain interventions, or brain prosthesis, is looked at as technological modifications specific to the brain. For example, the Hippocampus, which is largely responsible for turning short term memories into long term ones, could be replaced with a higher capacity electric organ. Bionic refers to physical enhancements through technology like prosthetic limbs. Exoskeleton refers to any internal enhancements or repairs to the spinal cord or major bone structures. Genetic modification refers to the process of editing genes at conception, and pharmaceuticals refers to ingesting medications and supplements that can enhance things like cognitive function (and even improve things like morality and emotional state).

This chapter will focus on economic implications spanning across all of these different types of human enhancement, but will take a primary focus on gene editing and wearable technology. This is simply because they are large enough topics to uncover on their own. Regardless of that, however, the most glaring commonality between all of these methods of human improvement is access to education, wealth and resources - a worry and anxiety many have as we move toward these applications. The previous arguments for these worries are flimsy at best. This chapter will address these concerns and discuss why these arguments do not stand up in reality. Before starting it should be noted that no one can accurately predict the future. Many of the assumptions and possibilities discussed are not guaranteed to take place, however for our discussion these assumptions are necessary.

Increased Disparity -
The Gap Between the Rich and the Poor

There has been much anxiety surrounding the possibility of enhancing genes in the human genome and its effects on inequality (this new technology was discovered by Jennifer A. Doudna for which she won the Nobel Prize in Chemistry in 2020). Some believe that this new technology in editing DNA will have a disproportionately positive outcome for the wealthy, and leave the unfortunate to fall further behind (which is an oversimplification of the issues). As written in the Frontier Technology Quarterly: Issue 2, May 2019 "the cost of gene therapies for rare diseases as approved in the United States and Europe can range from $373,000 to $1 million per patient per year. While genomics is shaping the future of medicine, the research is often targeted for certain population groups in mind, especially wealthy people who possess the ability to pay." (Scelta et al., 2019) Scelta et al. admit that "pharmaceutical firms in developed countries dominate genomic innovations, raising concerns of a "genomics divide" that can further exacerbate existing inequality in health outcomes between rich and poor nations." (Scelta et al., 2019)

It is no surprise that wealthy individuals would want to take every opportunity to increase their offspring's chances of success, as well as guarantee their well-being. There are certain traits that correlate with becoming a successful person. Those include traits like attractiveness, passion, confidence, athleticism, intelligence and creativity. While this is an incomplete list of a virtually infinite number of possible genes that could be manipulated, it illustrates that there are very real reasons for people to utilize this technology.

Another cause for concern is "the market demand for finding cures for rare diseases explains the rapid proliferation of gene therapies in the United States and other developed economies." (Scelta et al., 2019) While manipulating genes to make a person's chances at success more likely is a positive goal, however insignificant to the larger good, there is also the fact that this technology could be used to eliminate hereditary diseases from human beings at or before conception. If only wealthy individuals have access to gene editing technology, then they will be likely to live longer, giving them the opportunity to earn more wealth and gain more knowledge than those without access to the technology. This would likely further increase disparity between the rich and the poor. Heather Murphy of the New York Times states that "wealthy men and women generally have eight to nine more years of "disability-free" life after age 50 than poor people do, according to a new study of English and American adults." (Murphy, 2020) This is the alarming point: the rich are already highly advantaged over the poor in many areas of criterion. With the introduction and implementation of gene editing technology, the issue would likely get much worse.

Of course, up to this point, the discussion has been around reproductive enhancement - i.e. enhancing the genetic DNA of a human being before they are even conceived. The following sections will focus on human enhancement of people currently in existence. This is generally done through means of prosthetic limbs, surgeries, cognitive improvements and the like. This is sometimes referred to as human augmentation. Human augmentation devices include things like Apple watches and Fitbits - devices designed to help the user optimize their cognitive, physical, and mental health. The monitoring that these devices do on the heart, brain and body can indicate when a person should be resting, how a person is doing holistically, and what actions one should take to increase their health. These technologies are of course widely available, but generally speaking only the wealthier portions of the population use them.

Much of the anxiety surrounding gene-editing (Crispr technology) creating increased global wealth disparity is largely unwarranted. The scientists that use Crispr say "it is incredibly easy to use Crispr in the laboratory." And further on they give step by step instructions on how to use it on their website: "Though there are many methods to do so, the guide below will demonstrate how to select a gene to target, design a gRNA to delete that gene, and then actually delete it in living cells." (McCarty, 2018) The inventor of Crispr, Dr. Doudna, said in an interview that it takes about two weeks for her graduate students to figure out how to use it properly in the lab. The first case of using the technology on a human being was 2015, which was already 8 years ago at the time of this writing. The point is, it is not difficult at all to access this technology. There may still be plenty of work to be done in utilizing this technology to perfection, but it is likely going to be widely accessible to laboratories and hospitals without the need for extensive, specialized training. It would seem only a matter of time until it is common practice to edit the genes of human beings across the globe. Of course, the major argument and push by bioethicists is for a worldwide organization of laws and regulations to follow. The idea is that this wave of human enhancement is going to happen so quickly that we need to be prepared today with the considerations of tomorrow. This idea is further fleshed out in the coming chapters

Recent Economic Impact and Employment

Although the future implications of technologies like Crispr increasing the disparity of the rich and the poor are frightening, it is also a mere prediction - an assumption of where the world could be headed. J.M. Gitlin, from Calculating the economic impact of the *Human Genome Project*, says that "in 2010 alone, genomics supported more than 51,000 jobs, and indirectly supported more than 310,000 jobs, according to the Battelle study. This created $20 billion in personal income and added $67 billion to the U.S. economy." (Gitlin, 2013)

This study is 13 years old and the numbers have changed since then, but that might make it more important to note. This industry is not new, nor is it small. It has been generating money in the billions of dollars for a long time. This illustrates that a lot of wealthy individuals are interested in human enhancement. Anyone with a hereditary disease running in their family would be extremely interested in the technology. Nevermind the people that are inclined to ensure their offspring are given the utmost advantages in life. It is part of human nature to want the best for parents to want the best for their children. With this new technology that could mean that "competitive edge" could start before conception.

"A new report by research firm Battelle Technology Partnership Practice estimates that between 1988 and 2010, federal investment in genomic research generated an economic impact of $796 billion, which is impressive considering that Human Genome Project (HGP) spending between 1990-2003 amounted to $3.8 billion. This figure equates to a return on investment of 141:1." (Gitlin, 2013) The return on investment of this industry is most impressive. For every dollar that had been invested (as of 2013) there had been a return of $141 dollars. This is almost unheard of. The forecast for the industry is expected to grow, with estimates around $350 billion dollars USD by the year 2026.

Some numbers to note surrounding the size of the digital watch industry (i.e. fitbit and Apple Watch), the number of people working in these areas and the sheer amount of money being generated is staggering. "The global wearable technology market was valued at 61.30 billion in 2022 and is expected to expand at a compound annual growth rate of 14.6% from 2023 to 2030." (Grand View Research, 2022) That compound annual growth, if the prediction is correct, would put the market size at close to $400 billion dollars (USD). That is equal to one fifth of Canada's annual GDP!

While these technologies have made their way into common households, the applications for them are far-reaching. Hospitals and long term care centers are starting to utilize the technology to more easily track the health of their patients. Many of the data points of health are being covered in real-time, giving healthcare workers a more expansive list of information to work with on a second by second basis.

The single largest economic driver of wearable technology in the world is China. The enormous population in tandem with the lower costs of producing these items (relative to the west) are the factors being considered. Another point of consideration is that China manufactures almost all of these products. They are going to continue to grow their GDP until it surpasses the USA as the biggest economy in the world. Is wearable technology and bioengineering going to play a large role in the transition of China becoming the next superpower? It is hard to predict, as trends change. However, due to the uprising and correlation of wearable tech and GDP, the answer is likely yes.

Human Bioengineering, Genetic Disease, and its Impact on the Economy

The economy and the better path forward for humanity seems to come from the world's best and brightest. Most of the brilliant people like Galileo, Einstein, or Michelangelo all had a combination of inherited traits to make them the people they were (of course, they were also influenced by their surroundings, however, the debate of nature versus nurture is out of scope of this chapter. The previously mentioned individuals have all changed the course of history in some way with their influence, inventions and art - and most of the traits and characteristics they had might be identifiable at the level of the genome. These do not have to be psychological traits - they can be physical ones too.

Take for instance, the hippocampus. The hippocampus is an area of the brain responsible for converting short term memory into long term memory. This is one of the most vital mechanisms inside the brain for comprehending and retaining information. When trying to remember information for an exam, learn the steps needed for a recipe, or nail down the sequence of events of history, it is very important to have a highly functional hippocampus. If memories were not being converted from short term to long term, then knowledge would never be retained. Now imagine if the efficiency of the hippocampus could be improved upon, and allow anyone access to instant long term memory? Would the world see more brilliant, life-altering minds come into existence? What would that mean for the economy if traits like these were modifiable? Imagine employees that never forget anything! Now imagine other parts of the brain are modified to give a person improved concentration, problem solving skills, or logic and reasoning. Is the world far off from seeing super humans? The only limit to this technology is imagination (which might be another modifiable trait!)

When talking about healthcare issues in the United States, the periodical Health Affairs says that "rare diseases and conditions pose a tremendous collective cost burden to the country, likely in excess of $1 trillion annually." (Garrison et al., 2022) Now, rare diseases are just a sliver of the pie in hereditary diseases. Many heritable traits, like the likelihood to suffer from heart and stroke conditions, exist inside the human genome. Imagine a world where these parts of DNA are recognized and removed before conception. It would completely alter the landscape of our current healthcare systems globally. It would increase their influence and importance, as they would become a kind of "shop" where parents interested in having children could mold and modify their offspring.

Heart and stroke conditions are the most common form of death around the world. According to the University of Ottawa Heart Institute, "Many cardiac disorders can be inherited, including arrhythmias, congenital heart disease, cardiomyopathy, and high blood cholesterol." According to Dispatch Health, "many of the underlying conditions that can increase someone's chances of experiencing a stroke are hereditary." (Knowles, 2021) So while there may be no single trait that is the "stroke" trait or "heart attack" trait, there are many different combinations of genes that cause these conditions to exist. If the medical world can get proficient enough at figuring out these combinations, then it would be possible to remove them in advance of human existence. What would be left for healthcare professionals to do if everyone was pre-constructed with the healthiest, most advantageous DNA? As usual, new unforeseeable problems would likely arise.

Science has already identified many of these biological markers. For example, scientists have located genetic variations that create a protein that clears out arteries more efficiently than other proteins. The potential savings on healthcare costs would be astronomical in this one case alone. What if we could localize genetic variations that make us less likely to get sick, less likely to develop cancer, or less likely to become depressed or anxious? The applications of this technology are endless and that is why we must move forward with great caution. When we start talking about experimenting with genetic engineering, we are talking about experimenting with children prior to conception. The applications have great potential for economic value - more productive and healthy humans and lower healthcare costs due to decreased disease are just a couple of the potential benefits. But this is new territory to uncover, and there will likely be a number of unforeseeable negative outcomes. That is why this new area, however potentially beneficial for the economy, should be pursued thoughtfully.

Economic Positioning Going Forward for Biologically Enhanced Humans

With the amount of aforementioned uncertainties around ethical laws put into place, it is difficult to discern who is going to benefit most handsomely from genetic modification. Bioethicists around the world are of course crying for rules and regulations to be applied for all human beings, regardless of the country they find themselves in. If that becomes the case, then great. There will be an even playing field for all when it comes to Crispr technology.

The main concern is that the ability to utilize Crispr is apparently quite easy, and the enforcement of rules and regulations would be incredibly difficult to maintain. Think about criminal organizations. Do they care about the law? Clearly not, or they would be in a different line of work. Could the future become the wild west in terms of human enhancement? Also, what if not all countries agree to the worldwide terms? They could easily go off on their own and create superhumans to build up armies or to enhance their economy. The point to focus on is that this technology is not difficult to use. With easy access comes a whole swath of complications. Some are predictable but most are not. Creating realistic international standards is not an easy task, and to make things worse enforcing them would be nearly impossible. One of the current issues in biological science is "stem cell tourism" (Qaiser, 2020), where scientists and researchers move to countries with looser regulations to continue their work. The same problem is likely going to become prominent with human gene editing and engineering. To take this point one step further, consider a look at the potential uses of human engineering going on in China. NBC news reported in December of 2020 some harrowing words. "U.S. Intelligence shows that China has conducted 'human testing' on members of the People's Liberation Army in hope of developing soldiers with 'biologically enhanced capabilities..'" (Dilanian, 2020) Again, one must only use their imagination when it comes to the applications of these technologies. There is a potential that the least regulated countries could hold the most power, as they experiment and create advanced human beings to dominate their counterparts. This has implications far beyond just the economy.

Conclusion

It is likely that the economic impact and also the amount invested in human enhancement has only just begun to take shape. Currently, vast amounts of capital is being invested in researching ways humans can increase life expectancy, cognition, physical ability and the like (the actual list could go on ad infinitum). There is no limit to the potential impacts this could have on the world's economy, and a lot of it is going to take shape after bioethicists put forth rules and regulations. We may be a far cry away from the plot of Gattica playing out, or we may not be. We are in an incredibly precarious situation in regards to this topic, with a wide variety of players and political climates to contend with. One thing is for certain, however: the surface has not even been scratched to reveal the full implications of human enhancement on the economy.

References

"Are Strokes Hereditary?" DispatchHealth, 6 Dec. 2021, www. dispatchhealth.com/blog/are-strokes-hereditary/.

Dilanian, Ken. "China Has Done Human Testing to Create Biologically Enhanced Super Soldiers, Says Top U.S. Official." NBCNews.com, NBCUniversal News Group, 3 Dec. 2020, www.nbcnews.com/politics/national-security/china-has-done-human-testing-create-biologically-enhanced-super-soldiers-n1249914.

Garrison, Sheldon, et al. The Economic Burden of Rare Diseases: Quantifying the Sizeable ... www.healthaffairs.org/do/10.1377/forefront.20220128.987667.

Gitlin, Jonathan Max. "Calculating the Economic Impact of the Human Genome Project." Genome.gov, www.genome. gov/27544383/calculating-the-economic-impact-of-the-human-genome-project#:~:text=HGP%20produced%203.8%20million%20 job,%2463%2C700%20income%20per%20job%2Dyear.

"Inherited Cardiac Conditions (Genetic Disorders)." University of Ottawa Heart Institute, www.ottawaheart.ca/heart-condition/inherited-cardiac-conditions-genetic-disorders#:~:text=Many%20cardiac%20 disorders%20can%20be,indicating%20inherited%20genetic%20risk%20 factors.

McCarty, Niko. "CRISPR - How It Works, Top Applications and How to Use It Yourself." Medium, Medium, 15 Aug. 2020, nikomccarty. medium.com/almost-everything-you-should-know-about-crispr-how-it-works-top-applications-and-how-to-use-it-61e27b04bea6#:~:text=It%20 is%20incredibly%20simple%20to,delete%20it%20in%20living%20 cells.

MSc, Peter Joosten. "Human Enhancement Technologies: 7 Examples." Peter Joosten MSc., 9 Nov. 2022, www.peterjoosten.org/human-enhancement-technologies/#brain.

Murphy, Heather. "Rich People Don't Just Live Longer. They Also Get More Healthy Years." The New York Times, The New York Times, 16 Jan. 2020, www.nytimes.com/2020/01/16/science/rich-people-longer-life-study.html#:~:text=Just%20Live%20Longer.-,They%20Also%20 Get%20More%20Healthy%20Years.,of%20English%20and%20 American%20adults.

Qaiser, Farah. "Study: There Is No Country Where Heritable Human Genome Editing Is Permitted." Forbes, Forbes Magazine, 9 Nov. 2022, www.forbes.com/sites/farahqaiser/2020/10/31/study-there-is-no-country-where-heritable-human-genome-editing-is-permitted/?sh=2c452b8c7617.

Scelta, Gabe, et al. "Playing with Genes: The Good, the Bad and the Ugly." Frontier Technology Quarterly: Issue 2, May 2019, 1 May 2019.

Wearable Technology Market Share & Trends Report, 2030, www.grandviewresearch.com/industry-analysis/wearable-technology-market#:~:text=Report%20Overview,consumers%20is%20driving%20industry%20growth.

Chapter 9: Religion and Spirituality on Human Enhancement

By Iffah Shaikh

Introduction

Since the beginning of time, the influence of the major religions in the world has created and shaped a foundation of morals and values people uphold with high conviction – and beliefs surrounding human enhancement are certainly well outspoken in communities. The three Abrahamic religions: Islam, Christianity, and Judaism are widely interconnected as many core beliefs are the same and the believers worship one God; therefore, monotheism plays a large role in determining an organized set of beliefs, attitudes, and practices that center around submitting to the Creator of the universe, the All-Knowing, who advises to not succumb to worldly desires and temptations which inevitably cause harm to one's soul in the long-term (Kashiri, 2022). These rules are also put in place to ensure that whoever follows them, is guaranteed a great afterlife as this world is only temporary and a test (Kashiri). Major non-Abrahamic religions such as Hinduism and Buddhism are discussed in this chapter as they contribute unique and thoughtful perspectives on human enhancement because they integrate more of a spiritual approach of thinking. The question that remains now is, to what extent of human enhancement can society progress to and implement before it becomes an issue, and why?

Islam

Muslims follow their Holy book, called the Quran which is the word of God, and the teachings of the Prophet Muhammad PBUH from a collection of sayings called Hadith (Wiles). For Muslims, the answer to all of life's questions and phenomena is included in these books, including human enhancement. The views are not black and white, as it is very dependent on the type of situation regarding enhancement and if it is medically necessary to sustain one's life which is highly emphasized in Islam (Athar). Likewise, science and Islam are very much correlated with each other, and encouraged to explore. The Prophet states, "O, servant of God, seek remedies for those who have malady has already created its cure and its remedy" (Athar). For every ailment, there is a possible cure for it and humans should investigate and research them while reading the Quran to understand the basic foundations of the human body (Athar). In this case, human genetic modification done on somatic cells is permissible as the intention to modify these genes is to treat various medical disorders, if the purpose is therapeutic, it is recommended and permissible (Athar) However, there is definitely some gray area surrounding germline genetic engineering. There are many posed benefits for germline gene therapy, however with the current state of research there are far more significant risks of this process. The majority opinion from Islamic scholars is that this method of enhancement is prohibited if they would cause harmful consequences or unforeseen complications (Shabana, 386-411). But, if it is known for sure that it would cure a future illness that may cause a lifetime of difficulty for a baby, theoretically it would be allowed (Shabana, 386-411).

For non-therapeutic purposes, germline modification is highly disliked and the majority opinion is that it is definitely prohibited (Shabana, 386-411). The wisdom behind prohibiting human enhancement for cosmetic purposes is that human characteristics and facial features are created for each person in God's best form, and deemed as one's provision, or fate (Shabana, 386-411). Plastic surgery for cosmetic purposes is forbidden for this same reason, because for generations beauty standards and favored human characteristics have been subjective and are constantly changing; seeking validation from the creation rather than the Creator is highly discouraged and violates human dignity (Shabana, 386-411). Human cloning and fetal harvesting is also prohibited because this oversteps and disrupts the sanctity of marriage, the maintenance of lineage, and the unique creation of each human (Athar). Therefore, using donor sperm and eggs is also not allowed because if the sperm or egg comes from a person who is not the spouse, this may be akin to a certain degree of adultery because the Qur'an states: "And God has made for you mates and companions of your own nature and made for you out of them your sons, daughters and grandchildren" (Athar).

Essentially, there are three general guiding principles that can help mitigate which decisions are lawful to make regarding enhancement technologies according to Islamic jurisprudence: if something is prohibited (especially if it causes harm to the human body) but it becomes necessary for survival or if there is no alternate option available, if there are two harms one should accept the lesser of the two harms, and lastly, public interest will outcompete individual interests (Athar).

Christianity

Furthermore, Christian scholars who have specialized in biomedical sciences have examined biomedical moral enhancement through a religious lens. The views are very similar to Islam, as the moral compass and the preservation of human dignity are the main factors needing to be protected from the extremes of human enhancement applications. Human enhancement like AI, biomedical enhancement, and automation can undermine the practice of free will and using the will for good which is an inherent human action (Buttrey, 2). Specifically, one of the main theological concepts in Christianity is that the pursuit of life, and in order for life to flourish, must be accomplished by considering both the individual and the common good (Buttrey, 2). Hence, the Christian view on biomedical and human enhancement would factor in the question of how moral improvement or growth would be attained through the contributions of enhancements. One of the doubts that some Christian theologians have is whether or not enhancement technologies would make people arrogant or egocentric, which impedes on the likelihood of people encouraging community and cooperation (Buttrey, 11).

Protestant theologian Ted Peters explains that the Christian doctrine of imago Dei, "images of God" use human creativity as a privilege that people have the capacity to use for things like enhancement and therapeutic interventions that proclaim this ability of ours (Lustig, 81-88). In the same manner, Roman Catholicism also holds with confidence that human enhancement can be successful due to man's nature to do inherent good; it is trustworthy in the sense that humans are innately capable of being rational which is a result of God's grace in His creation, according to philosopher Thomas Aquinas (Lustig, 81-88). Despite sin existing, where humans can divert from their moral insight, men and women generally discern the good from bad and are inclined to do good and judge between intentions and actions(Lustig, 81-88). There is a mutual understanding of what is morally correct based on the truths in the Revelation of the Holy books and the innate innocence of humans; the Holy books, from the Old testament, being the true word of God provide a script of rules to distinguish between good and evil that humans come to terms with all by themselves. Therefore, based on these principles of Christianity does not prohibit human enhancement with respect to the therapeutic potential of it. However, there are definite constraints if the nature to use human enhancement is non-therapeutic. For instance, somatic cell in-vivo genetic engineering is allowed, but "therapeutic cloning" and germline gene therapy for cosmetic purposes is strictly prohibited because it tampers with and destroys embryonic human life (Lustig, 81-88).

Conservation of personal identity is emphasized in Christianity, everyone's soul is unique in singularity and placed into the body that is meant for them (Lustig, 81-88). According to Vatican Council II, the opinions and beliefs that drive ideas about human enhancement from a materialistic view is highly discouraged because the dignity of a person overpowers their biological state; Any changes that human enhancement causes for which is not necessary to a person's well-being is reductionist if it does not safeguard their identity, or "corpore et anima unus" (Lustig, 81-88).

On the other hand, due to the multitude of Christian denominations and sects, there are a group of people called Jehovah's Witnesses who belong to a branch of Christianity that have very differing views and interpretations of the Bible (Woolley, 715) . According to their belief system, when it comes to procedures like blood transfusions for example, it is prohibited because one must not sustain their life with someone else's blood and that abstaining from taking blood is obeying God's command and out of respect for Him being the Giver of Life (Woolley, 715). If blood products are not allowed, certain human enhancement technologies that are generally acceptable to the vast majority may be prohibited for Jehovah's Witnesses. The aim of this chapter is not to analyze these statements, but only emphasize that there are a range of opinions and understandings of the Biblical texts.

Judaism

The last of the Abrahamic faiths in this review, Judaism. Generally, the main sects of Judaism are reformist, orthodox, or conservative, and they are all in majority in support for the progression of human enhancement with favorable intentions (CCAR). For instance, according to the Central Conference of American Rabbis, they stated that they can not deduce a strict prohibition towards prenatal genetic diagnosis, or PGD for short (CCAR). PGD is a procedure where parents screen the genetic potential of zygotes and choose the most "desirable" fetus. Their rationale behind not forbidding this method is that PGD is not necessarily genetic engineering as no genes are individually being edited, but rather the parent chooses from a pool of existing zygotes (CCAR).

According to Jewish law, or the halacha, there are mixed views but also no strict prohibition on genetic modification (CCAR). Kilayim, one of the main concepts in Biblical literature is the mixing of species to breed animals among other things, is the basis that some scholars determine whether or not modification of genes is allowed (CCAR). The verse in Leviticus 19:19, states "You shall not let your cattle mate with a different kind," However, Jewish scholars mention that this popular verse can be misinterpreted and does not explicitly forbid the mixing and alteration of genetic materials in a lab, rather than mixed species mating together (CCAR). Other opinions, such as from the influential rabbi Nachmanides in the early centuries, mentions that God's creation is perfect and those who tamper or mix with the existing creation essentially deny or oppose God's creation (CCAR).

On the contrary, Nachamanides opinion has received contradicting views from many commentators of Jewish tradition and halakhic authorities – one opinion states that God has given humankind sovereignty over the earth to do as we please and essentially exploit and alter the universe for our own benefit due to God's trust in us (CCAR). So, some Jews have a more restrictive attitude while others have a very open attitude who base it off a quote from Genesis 1:28 "God said to [the man and the woman]: Be fertile, and increase, fill the earth and master it..."(CCAR).

Non-Abrahamic Religions and their View on Human Enhancement

Non-Abrahamic religions are also prominent as belief systems throughout the world. Buddhism, Hinduism, and Sikhism are a few of these larger non-Abrahamic religions. These religions have large and united followings and cultures which are distinct and navigate the topic of human enhancement a bit differently.

Buddhism

When delving into Buddhism, one will come across many schools of thought and traditions. However, the common belief that is taught in all Buddhist schools are the *Theravada* and *Mahayana* traditions (Hongladarom, 3) . These teachings state that ethics are determined based on the motivation behind a particular action (Hongladarom, 3). For instance, elongating the human lifespan would not necessarily be ethical or unethical without determining the nature of motivation behind it. Saving the human race from going extinct versus extending the lifespan for one to enjoy personal pleasures of life are two very different scenarios which are responsible for deeming an action wholesome or not (Hongladarom, 3). Many human enhancement technologies strive for humans to be stronger, smarter, and essentially immortal which stems from the human ego. Buddhist analyses state that if the ego wants to live longer simply for the duration of existence is unwholesome, but if the technology is used for altruistic purposes it can benefit both the individual and others as well (Hongladarom, 4). The difficult part is determining again, what constitutes an appropriate use of the technology for the intended purpose to help people. And, to what extent can enhancement divert from allowing humans to be capable beyond their normal capacities for it to then be unethical?

Principles of Buddhism state five important laws: utu niyama (physical law), bija niyama (biological law), kamma niyama (law of moral causation), citta niyama (law of mental activity), and dhamma niyama (natural spiritual law) (O'Brien). To briefly explain each law, physical law correlates with physics, geology, chemistry and explains the nature of non-living matter (O'Brien). Biological law is the opposite of the latter and touches on the laws of genetic inheritance and the life cycle of living matter (O'Brien). Kamma, or karma niyama is the law of moral causation and is a natural law that explains how one's thoughts, actions, and deeds have a "cause and effect" energy associated with them (O'Brien). Dhamma niyama is the law that explains the doctrines and teachings of Buddha. The law of mental activity; this is similar to kamma niyama but is more psychology based and explains the nature of the mind and one's state of consciousness (O'Brien). The reason for explaining these laws lays out the foundation of Buddhism – according to this philosophy, nothing is unnatural because everything that happens in the world follows one or more of these niyamas. With this logic, procedures like genetic engineering are not unethical because the research follows the natural law (Hongladarom, 7). Though everything that happens through these natural laws, the important thing to note here is the word "natural." If someone uses human enhancement technology to create an incredibly different, outlandish human in the future, this diverts from the traditional "normal" that is agreed upon in society (Hongladarom, 7). Therefore, the five laws can explain how things come into existence but do not imply if they are ethical or not, despite the action being natural or not.

Consequently, in *Dhammapada*, the most popular Buddhist scripture, has a passage that states, "Mind precedes all mental states. Mind is their chief; they are all mind-wrought. If with an impure mind a person speaks or acts suffering follows him like the wheel that follows the foot of the ox. Mind precedes all mental states. Mind is their chief; they are all mind-wrought. If with a pure mind a person speaks or acts, happiness follows him like his never-departing shadow" (Hongladarom, 9). A Buddhist would likely use this to shed some perspective on their process to judge whether or not a human enhancement technology would be of good use to implement in society. The meaning of this passage ties back to the point of motivation and intention behind an action; if the state of mind is pure or impure, for instance being selfish, filled with greed, or anger, then the act in itself will be impure and vice versa. An example stated by a Buddhist author Somparn Promta, is that if humans were to go to war and robo-soldiers were created where they are half human and half machine; they are stronger than their enemies, and can be used to save humankind from devastation (Hongladarom, 10). In this case, it would be a hypothetical scenario that would be considered ethical because there is an altruistic motivation behind creating a technology that would be more powerful than the ordinary human.

Hinduism

Hinduism is the third largest religion in the world, with Christianity and Islam being the first top two. It is one of the oldest monotheistic religions with very diverse and spiritual philosophies and practices (History). When it comes to the topic of human and genetic enhancement, there is no direct prohibition however the circumstances largely influence what would be considered ethical in Hinduism – a common pattern noted in most religions discussed so far. If an action has *daya*, or compassion behind it, this principle would make many therapeutic acts of enhancement permissible (Rajendran). As humans, we have a greater consciousness and power over the rest of living things and with this privilege comes great responsibility and virtue to ensure well-being is sustained in society and among all living things (Rajendran).

Way back in the Vedic period, originated three debts that every person owes when they are born (Rajendran). There is the debt to the celestials or the natural world, the sages who declared what the moral truth is, and to one's own family and ancestors. The importance of continuing one's lineage is emphasized, so increasing life expectancy and quality of life through gene therapy for example, serves the purpose of sustaining the continuity of family for future generations (Rajendran).

In regards to non-therapeutic purposes for human enhancement, this is where there is more of a hesitancy to allow certain technologies in Hinduism (Rajendran). One of the most basic concepts in Hinduism is *atman*, or the eternal spirit of a person. The soul is placed in the body which is merely a vessel for it to thrive in and grow. This spirit is reincarnated over and over again to learn and go through the journey to attain supreme blessedness (Rajendran). To achieve true personal transformation, the creative forces within a person are responsible to drive this change. Altering the physical and mental characteristics that one is born with breaches and contradicts the freedom of the spirit and delays the journey. Thus, true self improvement rather than body enhancement is valued in Hinduism because this is the improvement that leads to a greater awareness of God (Rajendran).

Moreover, another ethical principle central to Hindu belief is *ahimsa*, or no harm (Rajendran). Each human is a combination of physical, mental, and spiritual components and if a genetic change causes unintended consequences that may negatively affect the individual along with their offspring is highly risky; this potential for harm would deem certain technologies that are not fully researched prohibited due to *ahimsa* (Rajendran).

An interesting perspective on genetic cloning from a Hindu lens that many swamis, Hindu religious leaders, agree with is that there would theoretically be many issues in the future that cloned babies would face (Hinduism Today) . From an Ayurvedic point of view (Indian traditional medicine), a child born without vital life energy, or prana may be harmful to their life as they are conceived without love (Hinduism Today). Tirumantiram, an old scripture dating back to 2000 years ago, states that an embodied soul is highly influenced by the parent's energy. Other scholars state that clones may be a result of God's fate and the prevalence of a different species is due to His decree (Hinduism Today). Either way, these enhancement technologies are still very new and it is difficult to determine whether or not these manifested outcomes would be considered an act of "playing God" or not (Hinduism Today).

Mythology

One of the fascinating things about culture and society is the fact that human enhancement has been a desire for humans to explore and accomplish since the beginning of time – and while there are many religious doctrines and philosophies that are for and against certain implementations, mythology and folklore has proven that we have always admired and strived to be stronger, smarter, and more powerful versions of ourselves (Pew Research Center). For instance, in Greek mythology the skilled craftsman Daedalus created wings for his son Icarus and for himself. Similarly, Prometheus, the supreme trickster who defied the Greek gods would often steal fire and gift it to humanity (Pew Research Center).Turning back to Hinduism, there is also the tale of Raktabija, a mythological demon in Hindu literature who is a drop of blood, and he cloned himself every single time this drop of blood touched a battlefield (Hinduism Today). This would represent the concept of cloning and how it was discussed even in the ancient years. Fast forward to today, and human enhancement is still a hot topic with popular references to cinema culture showcasing Marvel movies with extraordinary superheroes that possess abilities that humans often dream of having as a figment of our imaginations, which most certainly exceed our natural capabilities.

Conclusion

In conclusion, it is evident that all of these religions have recurring themes of mutual agreement on the extent of human enhancement that would be considered ethical or not. There have been many debates and discussions from religious leaders all over the world trying to decipher through scripture and testimonies to analyze the relationship between human nature, desire, morality, and the effect on one's spirit if a particular technology were to be implemented. Human enhancement is certainly a complex topic that one must think critically about the future consequences of such promising yet foreign technologies to humans in this current period of time. Religions provide a wonderful basis of rational perspectives to help wrap one's mindset around abstract, unpredictable and new issues like human enhancement because they are often backed up by evidence and wisdom that people would not immediately think about as we are geared to seeing things from a very subjective lens.

References

Kashiri, Tendai. "Understanding Monotheism in Judaism, Christianity, and Islam." TheCollector, 2022, https://www.thecollector.com/understanding-monotheism-religions/.

Wiles, Stephen. "Research Guides: Islamic Law: Primary Sources - Qur'an and Hadith." Primary Sources - Qur'an and Hadith - Islamic Law - Research Guides at Harvard Library, 2020, https://guides.library.harvard.edu/c.php?g=309902&p=2070117.

Athar, Shahid. "Enhancement technologies and the person: an Islamic view." Journal of Law, Medicine & Ethics, vol. 36, no. 1, spring 2008, pp. 59+. Gale Academic OneFile, link.gale.com/apps/doc/A176480597/AONE?u=ocul_mcmaster&sid=googleScholar&xid=e6cbf9a7. Accessed 27 Feb. 2023.

Shabana, Ayman. Between Treatment and Enhancement: Islamic Discourses on the Boundaries ... 2022, https://onlinelibrary.wiley.com/doi/10.1111/jore.12404.

Buttrey, Michael, et al. "Faster, Higher, More Moral: Human Enhancement and Christianity." MDPI, Multidisciplinary Digital Publishing Institute, 13 Apr. 2022, https://www.mdpi.com/2077-1444/13/4/354.

Lustig, Andrew. "Are enhancement technologies "unnatural"? Musings on recent Christian conversations." American journal of medical genetics. Part C, Seminars in medical genetics vol. 151C,1 (2009): 81-8. doi:10.1002/ajmg.c.30197

Woolley, S. Children of Jehovah's Witnesses and adolescent Jehovah's Witnesses: what are their rights?Archives of Disease in Childhood 2005;90:715-719.

"NYP No. 5768.3." Central Conference of American Rabbis, 20 June 2018, https://www.ccarnet.org/ccar-responsa/nyp-no-5768-3/.

View of a Buddhist Perspective on Human Enhancement and Extension of Human Lifespan, http://www.assumptionjournal.au.edu/index.php/PrajnaVihara/article/view/1221/1074.

O'Brien, Barbara. "What Are the Five Niyamas in Buddhism?" Learn Religions, Learn Religions, 22 Feb. 2019, https://www.learnreligions.com/the-five-niyamas-449741.

"Hinduism - Origins, Facts & Beliefs - History." History, 2017, https://www.history.com/topics/religion/hinduism.

Rajendran, Abhilash. 2021,July 05 Monday. "Gene Therapy – Genetic Modification – How Hindu Religion Views Them?" Gene Therapy – Genetic Modification – How Hindu Religion Views Them?, https://www.hindu-blog.com/2021/07/gene-therapy-genetic-modification-how-hindu-religion-views-them.html.

Hinduism Today. "Playing God?" Hinduism Today, 1 June 1997, https://www.hinduismtoday.com/magazine/june-1997/1997-06-playing-god/.

Masci, David. "Human Enhancement." Pew Research Center Science & Society, Pew Research Center, 13 Apr. 2022, https://www.pewresearch.org/science/2016/07/26/human-enhancement-the-scientific-and-ethical-dimensions-of-striving-for-perfection/.

Chapter 10: Global Context of Human Enhancement

By Aidan Lang

The Global Context of Human Enhancement

The Current Landscape

Human enhancement and its implications are an incredibly complex set of topics when positioned within the global context. The narrative has largely been dominated by Western conceptualizations of self (Somech 161; Cousins 124) and medicine (Chui 30-56; Tseui 551). Further, the tenets of Western neoliberalism have allowed and will continue to allow for enhancement strategies to be pursued at a disproportionate rate (Sachdev 169). While some argue that these enhancement interventions will be beneficial to all (Buchanan1-34), others discredit this theory as narrow in scope (Sachdev 169-182). Critics of human enhancement (as the West defines it) argue that enhancement will have perilous consequences in the education (Wagner et al. 1-20) and labor (Dovidio & Gaertner 315-319) sectors. Most consequently, Western-led enhancement may add fuel to the proverbial imperialist fire—leading to rising international tensions (Desai et al. 489; Petras & Veltmeyer 1-10).

Values and doctrines underlining the day to day lives of individuals living across the globe vary greatly, and what one person defines as progress another defines as problematic. These values help to shape many of the institutions involved in human enhancement and its regulation; thus, a holistic understanding of human enhancement can not be achieved without positioning it within the global context. Further, it appears inevitable that implementation of human enhancement on a global scale would lead to an enhancement divide—an apartheid brought on by differential access to resources and willingness to incorporate enhancement within one's region (Buchanan 171-208; Sachdev 169-182). The first half of this chapter will cover the cultural differences between Western and non-Western regions that shape current ideas surrounding human enhancement; the second will discuss the logical implications of carrying out such ideas on a global scale.

Self concepts & identities

When seeking to understand human enhancement within a global context—or any emerging trends for that matter—there is a tendency for data to be skewed by insufficient representation (Hofstede 62). Various disciplines within the social sciences have been historically guilty of this when studying culture and identity. Only recently have social scientists begun to understand self concept and identity as geographically determined (Hofstede 43). Geert Hofstede's ground-breaking study on cultural conditioning suggests that theories of self and identity have hitherto been disproportionately routed in Western contexts (Hofstede 62)—a region that accounts for only twenty percent of the global population. Studies in social psychology link micro level concepts of self and identity with broader patterns of culture (Somech 161). There is extensive literature on the different manifestations of self based on geographic location. Individuals in Western, educated, individualistic, rich, and developed countries hold an independent view of self (Cousins 124). That is to say, the traits they use to describe themselves pertain little to external influences. When asked to describe themselves using the standardized Twenty Statements Test (TST), individuals from non-Westernized countries display a more interdependent view of self—placing emphasis on their social role over individualistic traits (Cousins 124). These varying definitions of self have been further studied by Markus et al. in their analysis of Olympic athletes who had recently won gold. In their study, Markus et al. compare and contrast the language used to describe Olympic athletes' performance on behalf of American and Japanese media outlets (Markus et al. 103). While both regions place emphasis on elite performance, results indicate a marked difference in how their performance was facilitated. American conceptualizations of agency tend to be both stable and disconnected from the influence of others. That is to say, an individual's unique set of traits are unchanging and entirely the product and possession of the individual (Markus et al. 110). Contrastingly, Japanese conceptualizations of agency were found to reflect a larger set of influences from both within and outside of the individual athlete—with past experiences, emotional state, and the actions/expectations of others in their social circle (Markus et al. 110). Where the identities of American athletes are their own, identities of Japanese athletes are far more interconnected (Markus et al. 110).

Markus et al. expertly highlights how cultural differences between Western and non-Western nations can manifest in different perceptions of identity and performance. This well documented phenomenon is important to understand as it relates to how human enhancement is conceptualized in and outside of the west. While Western ideas of enhancement reflect physical, cognitive, and emotional progress, these may not be universally defined in the same way, nor universally valued. This is reflected in differences in medical practices between Western and Eastern countries—a discipline that has dealt with human enhancement since before the term became a popular buzzword.

Health and Wellness

On a grand scale, there are several fundamental differences between Eastern and Western approaches to health and medicine. First, Western medicine separates health from illness, whereas Eastern approaches tend to see illness as an imbalance from a healthy state (Tseui 551). Second, Western approaches focus on altering one's environment where Eastern approaches place emphasis on adaptation to one's environment (Tseui 551). With different definitions of health and wellness come different methods of maintaining them.

In Western regions—specifically the United States—there have already been massive strides in human enhancement technologies that are reflective of individualistic self concepts. Pharmaceutical drugs are one notable example. Prescriptions of stimulants designed to manage ADHD medication have increased substantially over the past twenty years, and the current costs for ADHD treatments are estimated to surpass ten billion dollars per year in the United States (Graf et al. 1252). However, ADHD medication is only one potential treatment option. There are a wealth of therapies and family-centered approaches to treating ADHD that have often gone unstudied and underutilized due to Western medicines' conventional preference for pharmaceuticals (Schatz et al.

483). Conventionally speaking, Western medicine conceptualizes ADHD as an isolated illness with an isolated treatment.

Contrastingly, conventional Eastern wellness practices take a more holistic approach. Illness is not an isolated entity with an isolated treatment as seen in Western medicinal conventions (Tseui 551). Rather, a wealth of resources and avenues are taken into consideration when attempting to restore wellness in the East (Chiu et al 630). In their study of South and East Asian women with mental illness, Chiu et al. (630-656), assess the intersections of spirituality, gender, and culture in influencing their choice of treatment. Among the participants, South Asian women integrated many spiritual theories into their daily treatments. These theories were informed by Confucianism, Taoism, and Buddhism. The importance of religion and spirituality was professed by all but one woman in the study (Chiu et al. 650).

Chiu et al. show how Eastern cultures diagnose and treat illnesses as anything but isolated. While Eastern medicine focuses on restoring balance, Western practices show a clear orientation to "overcoming" illness. This is a major contributing factor to innovation in human enhancement technologies in the West. In addition to the West's preference for overcoming and getting ahead, the loose regulatory structures may further facilitate this phenomenon (Sachdev 169).

Regulatory Structures

The regulatory structures relevant to human enhancement also vary greatly by region. As previously cited, the debate surrounding human enhancement has largely been built on neoliberal ideals (Sachdev 169). There is no international consensus on "good governance" (Capps 252). While there is international consensus on managing risk and reducing harm to individuals, the tenets of Western neoliberalism allow for far less regulation and government oversight than would otherwise be seen in other regions. The 2018 Facebook data scandal perfectly exemplifies government regulators' inability to keep up with the innovation of the tech sector (Confessore). As such, it may be expected that human enhancement technologies will proliferate at a rapidly disproportionate rate across specific neoliberal areas such as the United States and the EU. This has already been observed in entrepreneurial endeavors such as Neuralink—an ambitious project aimed at creating an implantable brain-computer interface (Approach 2023). While many economically advantaged Eastern countries certainly have a vested interest in exploring forms of human enhancement, the programs are often state run. Placing large amounts of regulation and barriers to implementation.

The differential definitions and values underlying the debate of human enhancement can not be ignored. Already, there has been critique of the overwhelming individualistic traits of human enhancement technologies. The Global Futures Council has called for a revision of how human enhancement is conceptualized. Instead of focusing on individual benefit, the Global Futures Council advocates for equity and social cohesion when considering advancements in human enhancement (Baveilier et al. 204). Human enhancement should be reframed as collective welfarism—with the goal of benefiting not only the individual but the communities and societies they exist within (Baveilier et al. 204). Scholars posit that benefits to the individual can often be detrimental to the larger community; Bavelier et al. (204) use the example of performance enhancing drugs in professional athletes as well as cognitive enhancements in students. Bavelier et al. (204-205) also note that in ensuring autonomy and freedom in individuals' choice to benefit from human enhancement, they must not encroach on the autonomy and freedom of individuals who choose not to use enhancement. As can be seen, the cultural differences influencing the advancement of human enhancement are vast. Despite cultural differences, scholars, governments, and large organizations from around the globe have already shown both an extreme interest—as well as concern—for what the future of human enhancement will hold.

Implications for the Future

The proposed benefits of accelerating human enhancement appear vast. However, there are many questions raised surrounding the unintended consequences of such an endeavor. Allen Buchanan—a respected theorist on issues of biomedical enhancement—explores some of these topics in his book, "Beyond Humanity: The Ethics of Biomedical Enhancement". For Buchman, enhancement is inevitable and instead of arguing for or against it, one should seek to proceed in an informed and conscientious manner, lest the unintended consequences proliferate

(Buchanan 171-208). Crucially, the enhancement debate has been largely Western-centric, and if any nation is to undertake what Buchanan calls an "enhancement enterprise", previous patterns of inequality should be studied to better understand how enhancement may deepen these divides (Buchanan 1-34). This task may not be as abstract as it sounds; according to Buchanan, human enhancement is no different than any other form of enhancement, be it agricultural, industrial, or technological (Buchanan 39-44). By understanding how these phenomena have developed and affected nations across the world, a clearer picture of human enhancement's progression can be extrapolated.

Perspectives from Developing Countries

One potential negative consequence of accelerating human enhancement is a further marginalization of developing nations. Theorists posit that the debate surrounding human enhancement is inadequate as it pays little mind to the perspectives of these developing nations (Sachdev 169). While this chapter seeks to explore human enhancement in the global context, it is important to understand that the majority of the debate has hitherto been narrated by the West (Sachdev 169). Tenets of human enhancement are based on neoliberal ideals of minimal regulation and free enterprise (Sachdev 169). While Buchanan acknowledges that differential access to enhancement may periodically exacerbate inequality, he argues that the overall result will be progress for all (Buchman 1-34). Sachdev (172) critiques Buchanan's enthusiasm for progress as narrow in scope. Sachdev uses the example of the growing popularity of medical tourism in Thailand to highlight how supposedly benevolent Western interventions can have perilous consequences in the name of "progress" (Sachdev 177-178). Medical tourism promises economic benefit to developing nations, as individuals from developed countries travel to them to bypass rules or higher prices of surgeries in their home country. More facilities are subsequently opened to service these medical tourists. The argument being that more Western medical tourism equates to more readily available medical care for Thai citizens as well as greater economic strength. However, Thai medical tourism has, in reality, led to an uneven distribution of medical services. Doctors have transitioned from public hospitals to higher paying private clinics. In turn, hospitals raise wages to facilitate doctor retention, and funds that would have otherwise been allocated to direct treatment are lost

(Sachdev 177-178). By theoretically replacing these surrogacy clinics to human enhancement strategies, similar effects can be assumed.

Sachdev also speaks to the cultural impact of Western medical intervention in developing nations. Again, using the example of surrogacy tourism, Sachdev notes how carrying another person's child was originally frowned upon. Now, with the popularization of the procedure, public perceptions of surrogacy shifted to something virtuous (Sachdev 178). Some make the argument of normalization as "good". However, Sachdev notes that the imposing of Western views onto another nation is an example of post-industrial colonialism (Sachdev 178). Such interventions tend to localize economic regions within a developing nation—resulting in a growing economic division within the country (Sachdev 178-179).

Competition in Education

Cognitive enhancements have existed for many decades. There are widely popularized drugs to increase focus and attention (Schatz et al. 483). However, some scholars indicate that growing access and capability of cognitive enhancement may lead to a more competitive attitude in education (Wagner et al. 1). Wagner et al. recognize that inequitable access to cognition enhancing drugs will exacerbate inequality, however, the real issue is in the attitudes that cognitive enhancements promote (Wagner et al. 1). Cognitive enhancements place emphasis on competition over cooperation—getting ahead of one's peers instead of working with them (Wagner et al. 1). Again, the tenets of Western individualism are shown to take precedence in the debate, and Baveiler's (204) words of caution towards individualistic ideas of enhancement are echoed by Wagner et al.:

We should bear in mind that by changing our nature to fit society— rather than trying to create a society that fits our nature—we may inadvertently subjugate ourselves to the brute ideal of 'survival of the most competitive' at any sacrifice. Rather than encouraging a hyper-competitive society, we would do well to aspire and promote a more humanitarian society allowing for and accepting the limitations of innate human cognition. (Wagner et al. 14-15)

For Wagner et al. the very idea of cognitive enhancement appears to be built on Western individualism and competition. While these maxims are readily accessible to North America and Europe (Sachdev 169), they are not valued elsewhere (Cousins 124; Tseui 551). The effects of this are threefold. First, the West's disproportionate eagerness to "get ahead" may further educational divides (Wagner et al. 1). Second, the growing divide has the potential to force the hand of non-Western countries into adopting similar enhancement strategies—as cautioned against by Baveiler (204-205) and defined as a form of post-industrial colonialism by Sachdev (178). Finally, the inequitable utilization of human enhancements—cognitive or otherwise—allows for new forms of prejudice to proliferate. These prejudices may be most readily seen in disproportionate hiring rates of Western versus non-Western individuals (Dovidio & Gaertner 314-319).

Labor Market Discrimination

When developed, Western, and economically strong countries inevitably outclass developing countries in their access to cognitive enhancement capabilities, it may be presumed that individuals immigrating from those countries would face new forms of labor market discrimination. Implicit bias may lead to selective hiring practices in the labor market and contribute to growing economic divisions between people who come from countries who lack enhancement capabilities. There is already ample evidence towards the presence of implicit bias in hiring practices of organizations (Dovidio & Gaertner 315). In their systematic study, Dovidio and Gaertner uncovered that implicit bias greatly contributes to hiring rates of white versus black job applicants (Dovidio & Gaertner 314). These biases remain even in the presence of strict hiring equity policies and often manifest when candidates of different groups are equally matched on their resumes (Dovidio & Gaertner 315). Dovidio and Gaertner explain this phenomenon by positing that when all applicants are matched in their accomplishments, implicit bias may have room to operate in "indirect and rationalizable" ways (Dovidio & Gaertner 315). The differential access to human enhancement technology will add another layer of bias and misguided justification in the inequitable hiring practices of organizations everywhere. Applicants who are matched in all qualifications will always be subjected to unspoken understandings that Western-hailing individuals either have or grew up with more enhancement resources.

Ethical Polarization and Global Tensions

There have long been disputes over Western imperialism and post-industrial colonization of developing countries. Western NATO countries advocate for democratization while others label the process as repainted imperialism (Desai et al. 489). The introduction of human enhancement clinics in developing nations—as previously discussed (Sachdev 169-182)—may lead to accusations of colonialism and tensioning between nations. Surely, the enhancement apartheid led by the West holds the potential to destabilize many developing countries in the same way that military intervention has in South America and the Middle east (Petras & Veltmeyer 1-10). As discussed by Bavelier et al. (204-205), those who choose enhancement technologies must do so autonomously and freely, but in doing so, must not infringe on the freedom and autonomy of those who choose not to use them. This idea may be extrapolated to the level of international relations. In developing and normalizing human enhancement interventions, countries must not cause harm to those who either choose not to implement them or do not have the resources to do so.

Conclusion

In conclusion, the advancement of human enhancement is a highly complex, changing, and potentially divisive phenomenon in the global context. A narrative and mission largely dominated by the West has carried with it Western values, ideas, and goals. Fundamental concepts of identity and progress differ greatly between the East and the West (Somech 161; Cousins 124). These identities further inform definitions and approaches to health and wellness (Chui 30-56; Tseui 551), which in turn define approaches to their enhancement. Additionally, the regulatory structures of each region differ greatly. The Western free market has fostered a climate of unrestricted progress, competition, and individualistic gain (Sachdev 169).

If human enhancement continues on its trajectory, and if one is to extrapolate its implications from current Western-led enterprises, the picture appears to head towards a deepening of inequalities. Developing countries are subject to economic disparity under the guise of progress (Sachdev 177-178), education will be reframed as competition over resources (Wagner et al. 14-15), and the pre-existing biases that operate in the labor market will be strengthened by the enhancement divide (Dovidio & Gaertner 315). Finally, the enhancement divide holds the potential to further escalate pre-existing tensions over Western imperialism (Desai et al. 489; Petras & Veltmeyer 1-10).

References

"Approach." Neuralink, 23 Feb. 2023, https://neuralink.com/approach/

Bavelier, D., Savulescu, J., Fried, L.P. et al. "Rethinking human enhancement as collective welfarism." Nat Hum Behav, vol. 3, no. 1, 2019, pp. 204–206, https://doi.org/10.1038/s41562-019-0545-2

Buchanan, Allen E. Beyond Humanity? the Ethics of Biomedical Enhancement. Oxford University Press, 2011.

Capps, Benjamin J., et al. "Human Enhancement Technologies: Understanding Governance, Policies and Regulatory Structures in the Global Context." Asian Bioethics Review, vol. 4, no. 4, 2012, pp. 251–58.

Chiu, Lyren, et al. "Spirituality and treatment choices by South and East Asian women with serious mental illness." Transcultural Psychiatry, vol. 42, no. 4, 2005, pp. 630-656.

Cousins, Steven D. "Culture and Self-Perception in Japan and the United States." Journal of Personality and Social Psychology, vol. 56, no. 1, 1989, pp. 124–31, https://doi.org/10.1037/0022-3514.56.1.124.

Confessore, Nicholas. "Cambridge Analytica and Facebook: The Scandal and the Fallout So Far." The New York Times, 4 Apr. 2018, https://www.nytimes.com/2018/04/04/us/politics/cambridge-analytica-scandal-fallout.html

Desai, Radhika, et al. "The Conflict in Ukraine and Contemporary Imperialism." International Critical Thought, vol. 6, no. 4, 2016, pp. 489–512, https://doi.org/10.1080/21598282.2016.1242338.

Dovidio, John F., and Samuel L. Gaertner. "Aversive Racism and Selection Decisions: 1989 and 1999." Psychological Science, vol. 11, no. 4, 2000, pp. 315–19, https://doi.org/10.1111/1467-9280.00262.

Graf, William D., et al. "Pediatric Neuroenhancement: Ethical, Legal, Social, and Neurodevelopmental Implications." Neurology, vol. 80, no. 13, 2013, pp. 1251–60, https://doi.org/10.1212/WNL.0b013e318289703b.

Hofstede, Geert. "Motivation, Leadership, and Organization: Do American Theories Apply Abroad?" Organizational Dynamics, vol. 9, no. 1, 1980, pp. 42–63, https://doi.org/10.1016/0090-2616(80)90013-3.

Markus, Hazel Rose, et al. "Going for the Gold." Psychological Science, vol. 17, no. 2, 2006, pp. 103-112, https://doi.org/10.1111/j.1467-9280.2006.01672.x.

Petras, James, and Henry Veltmeyer. Power and Resistance: US Imperialism in Latin America. BRILL, 2016, https://doi.org/10.1163/9789004307421.

Sachdev, Vorathep. "'Beyond' Human Enhancement — Taking the Developing Country's Perspective Seriously." Asian Bioethics Review, vol. 14, no. 2, 2022, pp. 169–82, https://doi.org/10.1007/s41649-021-00193-z.

Schatz, Nicole K., et al. "Systematic Review of Patients' and Parents' Preferences for ADHD Treatment Options and Processes of Care." The Patient: Patient-Centered Outcomes Research, vol. 8, no. 6, 2015, pp. 483–97, https://doi.org/10.1007/s40271-015-0112-5.

Somech, Anit. "The Independent and the Interdependent Selves: Different Meanings in Different Cultures." International Journal of Intercultural Relations, vol. 24, no. 2, 2000, pp. 161–72, https://doi.org/10.1016/S0147-1767(99)00030-9.

Tseui, J. J. "Eastern and Western Approaches to Medicine." The Western Journal of Medicine, vol. 128, no. 6, 1978, pp. 551–557.

Wagner, Nils-Frederic, et al. "The Ethics of Neuroenhancement: Smart Drugs, Competition and Society." International Journal of Technoethics, vol. 6, no. 1, 2015, pp. 1–20, https://doi.org/10.4018/ijt.2015010101.

Conclusion

This book dealt with various issues of human enhancement. First, we outline what human enhancement is and the different technologies that constitute human enhancement. We outline how different types of human enhancement have contributed to human wellbeing. The history of human enhancement is introduced as well as the evolution from historic enhancement to today.

The discrete topics are then introduced and delve deeper into the focused issues of Human enhancement.

The second chapter tackles the new era of human enhancement and the various technologies that are being implemented at present day. Some of these technologies include: the field of genetic engineering, gene therapy, and new neurotechnologies. These technologies have the potential to permanently alter the paradigm and perspective of human enhancement for the coming generations.

The discussion then goes on to moral and ethical issues surrounding human enhancement in chapter three. We discussed the nuances of the ethical considerations related to this subject, including their stakeholders, context, and change over time. An overview of the major relevant ethical considerations such as freedom and autonomy, fairness and equity, and human dignity and identity was also provided. Lastly, to conclude the discussion, the aforementioned ethical considerations were applied to a case study concerning pharmaceutical cognitive enhancements.

Chapter four explores the social and political implications that human enhancement can have on both society and the medical world. While there are evident biological effects of human enhancement, much of policy and the general reality of the human condition will be affected by this significant strive towards new medical technologies. Similar to the means through which medicine is scrutinized and regulated today, we see that there are broad implications that human enhancement can have on policy making spaces that will likely be bipartisan in their support. Similarly, society will have broad and diverse considerations into how this mode of potentially manipulating the human condition can affect humanity in the long term and the overall shifts that human enhancement can have on the way that individuals live their life when a part of the collective body.

Following this, in chapter five, the philosophical questions are explored. The debate of ethics, social justice, and the impact on human relationships are all things to consider. By looking at different philosophical perspectives that have guided us for decades, we can further understand how different people view, understand, and react to this subject. We will explored various different philosophical perspectives such as utilitarianism, deontology, eudaimonism, existentialism, and both ethics of care and virtue ethics. These topics were all discussed in view of human enhancement.

The psychological issues of human enhancement were tackled in chapter six. A unique consideration for the psychosocial impact of human enhancement in the existing vulnerabilities in those who are drawn to and seek elective cosmetic procedures and surgery. Ultimately, this chapter serves to explore the need for greater psychosocial consideration in the process of elective human enhancement procedures

Chapter seven deals with the regulatory and legal aspects and challenges of Human Enhancement. Human enhancement if left unregulated poses significant issues for society and the concept of what it means to be human. Thus, many governments and agencies have attempted to regulate the use of human enhancement; however, the attempt to implement regulatory and legal frameworks comes with many issues that lawmakers, ethicists, and agencies must first overcome.

Chapter eight deals with economic issues of human enhancement. The chapter focused on genetic modification and wearable technology. The chapter built on the economic principles surrounding developing human enhancement technology, and introduced these concepts to provoke thought about the economic challenges that come with this technology.

Chapter nine then dealt with the spiritual and religious aspects of human enhancement. While investigating this topic, key beliefs that some major religions of the world follow and preach and many similarities were found despite how different the practices of each religion are. Lastly, the chapter concludes by discussing how cultural influences like mythology and cultural customs have shaped and created the foundation for humans wanting enhancement technologies since the beginning of time.

Chapter ten then embarks on the discussion of globalization and how global accounts are taken into consideration. The values, perceptions, and ideologies informing current practices of human enhancement hold the potential to facilitate an enhancement divide between Western and non-Western regions. If the enhancement debate continues on its Wester-centric path, current inequalities may be exacerbated.

The book is used as an introductory source of human enhancement issues. These chapters are meant to provoke thoughtful discussion and bring to attention the various issues surrounding the subject. Although there are many specialized topics and other issues and ideas which were not included, the intention is to use the information found here to catapult the thought process around these new topics.